Social Futures, Global Visions

The United Nations Research Institute for Social Development (UNRISD) is an autonomous agency that engages in multi-disciplinary research on the social dimensions of contemporary problems affecting development. Its work is guided by the conviction that, for effective development policies to be formulated, an understanding of the social and political context is crucial. The Institute attempts to provide governments, development agencies, grassroots organizations and scholars with a better understanding of how development policies and processes of economic, social and environmental change affect different social groups. Working through an extensive network of national research centres, UNRISD aims to promote original research and strengthen research capacity in developing countries.

Current research themes focus on the social dimensions of economic restructuring, environmental deterioration and conservation, ethnic conflict, the illicit narcotic drugs trade and drug control policies, political violence, the mass voluntary return of refugees, and the reconstruction of wartorn societies, as well as ways in integrating gender issues into development planning.

Social Futures, Global Visions

Edited by
Cynthia Hewitt de Alcántara

Blackwell Publishers Ltd. and UNRISD

Blackwell Publishers
108 Cowley Road, Oxford OX4 1JF, UK
and
238 Main Street, Cambridge, MA 02142, USA

British Library Cataloguing in Publication Date
A CIP catalogue record for this book is available from the British Library

Library in Congress Cataloging-in-Publication Data

Social futures, global visions/edited by Cynthia de Alcámtara. p. cm.
"This text was originally published as special issue 27:2 of Development and Change"—T.p. verso.
Includes bibliographical references and index.
ISBN 0-631-20230-7 (alk. paper).—ISBN 0-631-20229-3 (alk. paper)
1. Social prediction. 2. Social change. 3. Social history.

I. Hewitt de Alcántara, Cynthia.
HN17.5.S594 1996
303.49–dc20
96-7826
CIP

This text was originally published as
special issue 27:2 of
Development and Change,
published by Blackwell Publishers Ltd.
on behalf of the Institute of
Social Studies, The Hague

Printed in Great Britain by Whitstable Litho, Whitstable, Kent.

Contents

Preface

During a weekend in March 1995, more than a thousand participants from all over the world gathered in the Banquet Hall of the University of Copenhagen to debate the great social issues of the closing years of the millennium. Less than a kilometre away, 120 heads of state and governments assembled at the same time to deliberate upon and adopt a set of commitments and a programme of action for global social development — a Social Charter for the twenty-first century. The occasion for these events was the World Summit for Social Development — a striking testimony to growing alarm over deteriorating social problems around the world and a spectacular attempt to restore primacy to social concerns in development policy.

Taking advantage of this extraordinary occasion, the United Nations Research Institute for Social Development (UNRISD) invited a group of distinguished writers and social analysts to reflect on the fundamental processes driving social change in the closing years of the century. The premise underlying this endeavour was that the scale and speed of change in recent years have made existing paradigms and models increasingly less relevant to understanding the nature of contemporary social dilemmas and that new thinking is required to provide more appropriate conceptual and institutional frameworks for coping with escalating social problems.

The world has been convulsed by momentous changes over the past two decades. The global revolution in economic policy and organization is sweeping away internal and external barriers to the sway of free markets and private enterprise. The most dramatic manifestation of this process is the dismantling of communism in Europe, thus heralding the end of the Cold War — the dominating influence in world politics and economics for nearly five decades. Such changes have given fresh impetus to economic globalization, further buttressed by astounding advances in technology. Social and cultural domains have been profoundly modified by the global reach of travel, communications, media and consumer values. The doctrine, if not always the practice, of liberal democracy and human rights has become an integral element in the process of globalization.

Changes of this order of magnitude and speed cannot but transform the universe of social problems and the technological, economic and political environment for devising social policies. The values and identity of individuals, as well as their sense of security, are challenged; institutions like the family and community are subjected to intense stress. New patterns of personal relations, social support and organizations emerge. As forces binding people into nation states begin to crumble, critical questions are raised about their purpose and legitimacy. The emerging consciousness of a global community is set against more parochial visions.

The essays presented in this collection are an attempt to interpret and illuminate the fast-moving kaleidoscope of social changes ushered in by the forces of globalization. They have in common their concern with the impact of these forces on human welfare and solidarity. It is hoped that they will spur further reflection and action on policy reform and institutional renovation, and that they will promote social development in our rapidly changing world.

Dharam Ghai
Director, UNRISD

Acknowledgements

The UNRISD conference on 'Rethinking Social Development', held in Copenhagen on 11–12 March 1995, was a collective effort based on the contributions of many people. The Institute is grateful, in the first instance, to the group of ten speakers whose papers set the stage for debate. It is also indebted to the University of Copenhagen (represented by its rector, Kjeld Mollgaard) for providing a beautiful venue and to Birgitte Sorensen, who served as liaison between UNRISD and the Danish institutions involved in preparations for the Social Summit.

Numerous colleagues at UNRISD formed part of the early planning group for the conference and gave valuable comments on its organisation. This group included Dharam Ghai, Yusuf Bangura, Solon Barraclough, Peter Utting, Jessica Vivian, David Westendorff and Marshall Wolfe. The death of Wolfe in January 1995 was a terrible loss to the group, and to the development community.

Arrangements for the event itself were ably handled by Adrienne Cruz, Jo Grin-Yates, Véronique Martinez, Irene Ruiz de Budavari and Wendy Salvo. Jenifer Freedman provided efficient editorial assistance.

The conference book, *Social Futures, Global Visions*, is the result of collaboration between UNRISD and *Development and Change*. The Institute would like to express its sincere appreciation to the Board of Editors of that journal (Martin Doornbos, Ashwani Saith and Ben White) and to the Assistant Editor (Paula Bownas) for their invaluable help.

The programme on 'Rethinking Social Development' was carried out with the support of the United Nations Development Programme (UNDP). The opinions expressed in the following essays are those of the authors themselves, and publication does not necessarily constitute an endorsement by UNRISD or any other part of the United Nations system.

Introduction

Cynthia Hewitt de Alcántara

I

Future generations, looking back on the last years of the twentieth century, will see a contradictory picture of great promise and equally great uncertainty. The 1990s have all the symptoms of a 'turning point' in world history — a moment when many of the structural 'givens' of social development themselves become problematic and world society undergoes profound reorganization.

Throughout broad areas of the world, the very nature of the economic and political system is a subject of contention. Visions of a better world are changing rapidly, as individual initiative takes precedence over collective interests and communist utopias give way to capitalist dreams. At the same time, the centre of the international economy is shifting from industrial to service sectors, and from trade to finance. Transnational conglomerates manage an increasing share of world wealth. The gap between rich and poor is widening markedly, and more and more people live out their lives in abject poverty. Nations are riven by conflict based on religious or cultural differences which, it was previously supposed, would be lessened rather than sharpened with the passage of time.

These developments occur within a framework of rapidly expanding social and economic interdependence on a global scale. Communications satellites and mass transport increasingly bring almost every human being on the planet within reach of every other. Computers manage the virtually instantaneous flow of information among individuals and groups who pursue common ends without ever having met. These are powerful forces for cultural integration, promoting common interests, values and aspirations among hundreds of millions of people.

Technological innovation has also played a fundamental role in encouraging the remarkable increase in world trade — and the even more remarkable internationalization of finance — over the past few decades. National, subnational and local economies are now linked in extremely complex networks which are as geographically extensive as they are inherently fragile. The commitment to deregulation evinced by most governments in the 1970s and/ or 1980s has reinforced the extent of global economic interdependence, as well as its fragility.

Obviously, expanding the boundaries of exchange and cultural contact creates both opportunity and risk. Human history has been played out against a background of expanding contacts among peoples, in which the life chances of some groups are improved and those of others are devastated. In this sense, growing social and economic integration always gives rise to conflict. But the latter is particularly acute in hard times. Although some areas of the world have seen remarkable growth during recent decades, the rapid elimination of cultural and economic barriers between nations and peoples has proceeded for the past twenty years within a context of stubborn and recurrent recession, as well as increasing indebtedness over a considerable part of the developed and the developing world. In consequence, individuals and firms in both North and South have been forced to develop new strategies of economic survival within a particularly hostile environment, in which much larger numbers of competitors vie for a relatively static — and in some cases a dramatically shrinking — pool of resources.

Pressures on the labour market, the commodity market and the business community are severe under these circumstances. Even in relatively traditional rural areas of the South, for example, small farmers long producing agricultural commodities for export find themselves competing with large numbers of similar producers in other parts of the world, recently drawn out of subsistence production by their own governments' programmes to encourage exports. Unskilled labourers throughout the Third World must also seek work in a market where the number of available hands increases under the impact of expanding monetary relations and newly induced requirements for consumption. Skilled workers in all countries confront the challenges posed by new forms of organization of industrial production, which increase efficiency through subcontracting and the more flexible use of labour.

Competition for jobs and markets has been dramatically heightened by the collapse of barriers formerly separating the Soviet Union and Eastern Europe, as well as China, from the industrialized West. These represent vast new pools of cheap and often relatively well-trained labour, entering the world market at a time when competition among enterprises in industrialized and developing countries alike depends to an increasing extent upon ensuring that firms have access to a low-cost workforce.

Up to now, the livelihood of workers in the industrialized world, as well as in many newly industrializing countries, has to some degree been protected by labour agreements and broader political pacts hammered out within the boundaries of the nation state. Nevertheless the liberalization of trade, combined with deregulation of finance, has weakened or destroyed barriers which once retained many economic activities within national boundaries. In consequence, employers and investors are now less constrained to bargain with workers within their own countries than was previously the case.

At the same time, national governments have been forced by liberalization and deregulation to strengthen the competitive position of their economies in the global arena, and thus to adopt measures which attract foreign capital and cheapen exports, even when these measures may threaten the standard of living of large numbers of people and the capacity of the government itself to meet its obligations to citizens and to the environment.

The declining capacity of most national governments to provide social services constitutes an additional element in the progressive disprotection of many groups which could formerly count on certain minimum social benefits. The most dramatic changes in this area have no doubt occurred in Eastern Europe and the former Soviet Union, where a sharp decline in living standards has been associated with the disintegration of the socio-economic and political system. But the sense of growing insecurity is hardly less throughout much of the Third World, burdened by heavy debt, and among the many millions of people in countries like China, where economic liberalization has encouraged both rapid growth and the loss of entitlement to existing forms of social protection for those who migrate in search of opportunity.

Over the past few decades, migration has become a central element in the livelihood strategy of more and more people around the world. Although current flows of people in search of a better life may be proportionally no larger than they were at certain times in the nineteenth century, they involve far greater diversity of cultural contact and very large absolute numbers of migrants. The potential for disruption of existing forms of social organization in the communities and regions of origin of migrants is sometimes great — particularly when most able-bodied members of households depart, leaving the young and the old to cope as best they can. Women who remain behind when men migrate must assume new roles and add new tasks to those traditionally assigned them. The potential for improving the standard of living of migrants' families is also considerable, as remittances are sent home and invested. Some migrants get ahead, and some find departure from their place of origin a form of liberation from oppressive obligations. In all too many instances, however, migration remains a harsh necessity — a last resort involving privation and not infrequently the danger of physical harm.

In major receiving countries, international migration creates enormous problems of social integration and cultural adaptation which are currently at the centre of the policy debate. The juxtaposition of people who often share neither a common language nor a common religion, and who have very different customs, makes unusual demands on human tolerance and understanding. The arrival of large numbers of foreigners also creates unusual strains on existing social services and local economies.

As global integration quickens at the end of the twentieth century, barriers between different cultures are of course being torn down — and relations between different cultures strained — whether migration occurs or not. Even if people do not leave their homes, local identities may be challenged by what is broadcast on television and radio, or seen on film. The revolution in mass communication, which encourages the creation of a global culture, has great potential for promoting understanding and solidarity, and enhancing knowledge throughout the world. In its present form, it also has an awesome capacity to exalt consumerism and proffer a highly individualistic definition of a good life.

To attain this good life is not easy in hard times, whether one lives in a developing or an industrialized society. Competition among those who struggle to be included in the global consumer culture creates sharp rifts in virtually all communities, from rural India to midtown New York. Local forms of solidarity are replaced by new values and ties, which link small groups with access to the new style of life in one city or country to others within the region and the world — while increasing the gap between these minorities and the majority of the inhabitants of the planet, who remain excluded from the consumer society they observe from afar.

In a large number of cases, illicit and illegal activities constitute one of the few avenues currently open to many aspirants for inclusion in the global consumer culture. Particularly in situations of wide-ranging social reorganization, such as those prevailing in many socialist or formerly socialist countries, the definition of what is illicit and what is not tends to be unclear, and activities are tolerated which in previous periods would have been morally unacceptable. At a time when many people confront narrowing economic opportunity, participation in the illegal economy furthermore constitutes one of the few realistic options for ensuring a basic level of subsistence. Illegality makes certain commodities or services unusually profitable. Thus the drug trade has become one of the central economic activities of the late twentieth century, drawing millions of people — from the peasant villages of Third World countries to the inner cities of the industrialized North — into networks of exchange which provide great wealth for some and a tolerable living for many who have limited alternative sources of income.

Needless to say, both the dangers and the opportunities inherent in this period of expanding international integration are heightened by the end of the Cold War. The collapse of the Soviet bloc, which is itself a stunning manifestation of the force of global integration, eliminates one of the central organizing principles of post-war international society and encourages a revival of animosity among peoples of different cultures who were formerly required to live together within multi-ethnic states. Concurrently, it creates new opportunities for peace-making in countries where competition among the great powers long dictated support for antagonists in internal wars.

Ethnic conflict, and the degeneration of ethnic conflict into civil war, is one of the unfortunate hallmarks of the late twentieth century, as it has been during earlier periods of world history marked by a profound reorganization of social and political relations. This form of conflict rests on reaffirmation of an exclusive form of identity and solidarity among people who share common historical grievances and have defined a common enemy. Although differences of religion and language are often involved, this is not necessarily the case. In fact, personalistic factions can manipulate enmity in ways which create self-sustaining spirals of violence among antagonists.

At present, we are witnessing dozens of civil wars across the globe, with hundreds of thousands of casualties. There could be no more telling sign of social malaise than the atrocities committed in these wars, usually against civilians. Yet while international organizations and groups search for ways to deal with such occurrences, violence is supported and even encouraged by long-standing global institutions, including the arms trade (which makes the instruments of death available) and sectors of the film industry (which both trivialize and glorify sadistic forms of violence).

Rapid economic and social reorganization, accompanied by far-reaching cultural change, makes unusual demands on political institutions. Economic uncertainty and fear of marginalization encourage electorates in established democracies to favour immediate remedies over long-term policies; and the same fears immensely complicate the task of creating effective democratic regimes in countries where such systems of government are only now being established. Furthermore, the global nature of so many of the problems of today reinforces the need for a far more effective system of international governance than that currently available. There is a striking incongruence between patterns of social integration which bind people around the world more closely together than ever before, on the one hand, and the frailty of existing mechanisms for discussing joint problems and promoting joint action on the other.

II

In the opening essay of this volume, Ralf Dahrendorf considers the constraints and choices faced by governments in the advanced industrial world as they confront the imperative of international economic competition. He reminds us that there *are* options and that capitalist economies are anything but uniform in their approach to the problem. Some choose to compete through implementing a low-wage strategy while others emphasize high skills. Some rely upon low taxes and high profits, while others prefer high taxes and low distributed profits. Protectionism is certainly practised. Nevertheless, 'these are minor variations on the major theme of globalization', with its attendant pressure for ever-greater 'flexibility' and personal insecurity.

Heightened global competition is exacerbating a trend toward social exclusion within OECD countries. Dahrendorf is not opposed to the maintenance of some degree of inequality within society, but he is very concerned by what he describes as 'inequalization'. There is a systematic divergence in the life chances of large social groups in advanced industrial nations and an increase in the proportion of the population which is 'simply not needed' within the economy. 'The rich can get richer without them; governments can even get re-elected without their votes; and GNP can rise and rise and rise'. This process enlarges the mass of 'strangers' within OECD societies, made up not only of the economically unnecessary but also of the culturally different. Dahrendorf is concerned above all with the damage which such polarization can inflict upon social cohesion and political freedom in the developed world. Growing insecurity gives rise to intolerance, xenophobia and a more general temptation to consider authoritarian solutions to problems of law and order; and the marginalization of increasing numbers of people weakens the very basis of citizenship.

After considering a series of domestic policies which might improve the life chances of people in the OECD countries and protect the legacy of democracy during troubled times, Dahrendorf tables the unanswered question of how to design international responses to the challenge of globalization. As he notes in closing, 'We are talking about prosperity for all, civil society everywhere, and political freedom wherever people live. This means that in the end we are not concerned with privileged regions, but with one world and its appropriate institutions'.

In the following essay, Fatema Mernissi looks at that world from the vantage point of the Middle East and North Africa, describing the impact of international forces on the population of a region which occupies a notably different geopolitical position from that of the OECD countries. For a person in the Arab world, exposure to the harsh effects of international economic and political competition has for many years been more direct and threatening than it has been for most citizens of advanced industrial nations. Authoritarianism, and not democracy, has been centrally related to the requirements of participation in the international market for oil.

Mernissi invites us to understand the anger and despair of millions of unemployed and impoverished inhabitants of the Arab world whose life chances are worsened by authoritarianism within their countries and geopolitics outside them. The fate of these people is linked in tragic and perverse ways to the well-being of citizens of OECD countries. The very stability of international oil markets, on which the economic security of advanced industrial economies is founded, has been ensured through weaving alliances with interests (both public and private) in oil-producing countries which are often neither politically accountable nor able (or in some cases, willing) to use oil revenues to promote workable development alternatives. All too often, the profits obtained from oil in the Middle East have simply been returned to the West to pay for arms.

In other words, as Mernissi puts it, ensuring 'certainty' in some parts of the world system creates enormous 'uncertainty' for other parts of it — and the situation may well be worsening under the impact of increasing global competition for investment and employment. Important sectors of developed country economies depend upon military industries, whose interests are promoted by Western governments. Employment generated in France or the United States by the military purchases of Arab governments will not be generated, to the benefit of Arab citizens, in Egypt or Saudi Arabia.

Mernissi, like Dahrendorf, is searching for ways to promote tolerance and democracy. Within the Arab world, this requires dealing with Islamic fundamentalism; here Mernissi reminds us that encouraging a rational, secular civic discourse in the Middle East and North Africa depends not only upon the commitment and sacrifice of the local population, but also upon a change of policy on the part of Western democracies which have a history of supporting fundamentalists as a bulwark against communism or Arab nationalism.

In the light of this long record of Western support for Islamic fundamentalism, the currently fashionable intellectual debate on the 'clash of civilizations' (in which a rational, democratic West is threatened by an irrational, authoritarian East) seems frivolous at best. Mernissi underscores the point that many people within and without the Muslim world have a common interest in promoting peace and justice in the Middle East and North Africa. 'In fact, the most difficult of all intellectual enterprises today is the attempt to separate "East" and "West" '.

III

Are nation states — whether in the East or the West, the North or the South — the best institutions for dealing with the enormous challenges thrown up by global technological and social change at the end of the twentieth century? The fact that such a question could be asked at all is a sign of our times. And it is being asked, not only by those who would hope for a more integrated regional or global response to the problems at hand, but also by others who espouse a world of autonomous individuals making rational choices without structural constraints of any kind.

Eric Hobsbawm reviews the basic attributes of territorial states, as these have developed over two centuries or more, and analyses the challenges they face. Like Dahrendorf, Hobsbawm is troubled by the relatively declining capacity of Western liberal democracies to protect their citizens in a globalizing international context, and by the attenuation of links between citizens and public affairs. The latter weakens the general commitment to the common good and reduces the capacity of national governments to redistribute resources from the least to the most vulnerable members of society.

Among other things, Hobsbawm notes, well-established nation states (in whatever part of the world) are structures of solidarity, the broadest manifestation of mutual obligation and support in a set of relations beginning with the family and the community. The current liberal critique of government legitimizes an attempt on the part of better-off segments of society to opt out of this pact of solidarity. The same can be said of numerous efforts to decentralize, privatize or secede from existing states: prosperous regions are usually much more likely to promote political fragmentation than poor ones.

Therefore, while reform is no doubt necessary in many areas of public administration and government, Hobsbawm recommends prudence. Neither ultra-liberalism (which destroys the state from within) nor fragmentation (which reorders its boundaries and jurisdictions) is a panacea for the serious problems afflicting all modern states. On the contrary: 'Insofar as the problems of states depend on size, they reflect the fact that states are too small to cope effectively with the global, transnational and certainly supranational scale of the world'. Furthermore, the quality of democracy is not determined by the physical distance between the governors and the governed, but by the effectiveness of public institutions. A world 'composed of nothing but states of the size of Luxembourg or Antigua' would offer fewer opportunities for the exercise of democracy than one composed of larger states like France and the United States, because in the former case the decisions affecting citizens' lives would be taken almost entirely by 'non-state entities beyond the control of any voters'.

In sum, until supranational authorities can provide an effective space for democracy and redistribution, Hobsbawm concludes that the nation state is 'still the best unit we have for the time being'. He reminds us of the 'catastrophes into which a sudden and uncompensated collapse of the state, as in the ex-Soviet Union, can plunge a large part of the world'.

A second perspective on the future of the state emerges in the following essay, by the Nobel Prize-winning novelist Wole Soyinka. The 'crisis of nation-being' in Africa is of a different nature and order than that experienced by citizens of Western democracies. In fact, while the inhabitants of advanced industrial societies discuss alternatives for reforming their relatively stable administrative and political systems, large numbers of people in Africa survive in situations of great instability and fear. Soyinka surveys the 'dismal picture of a continent that continues to smile at its image in the mirror while the whole world looks on in tears'. After drawing up a damning inventory of poverty, hunger, human rights abuses and war, he asks what basic criteria must be met in order to 'address a populated space as a nation'. His answer is clear: the human beings within that space must enjoy certain basic rights, the government must be founded on the will of the people, and there must be a collective effort to provide a minimally decent livelihood to all.

On the basis of these criteria, he questions the validity of applying the concept of statehood in many parts of Africa. The political entities created during the process of decolonization on that continent have, in his opinion, too often proved to be nothing more than 'a gambling space for the opportunism and adventurism of power'. Rapacious and corrupt regimes find it convenient to hide behind the façade of 'nation-being', which enables them to siphon resources from the bottom toward the top (a different kind of redistribution from that discussed above) and to enjoy a false legitimacy in international society.

The metaphor which Soyinka chooses to represent a functioning form of national solidarity is 'My father's house' with 'many mansions', including all the rooms of clans and ethnic groupings. When personalistic factions abuse their power in that collective space, the house, divided against itself, cannot stand. People withdraw into separate compounds, building their identity and strength increasingly on ethnic roots, and the possibility of ethnic violence is heightened. This is, unfortunately, the future which Soyinka foresees for Nigeria, if the current authoritarian regime remains in power.

While Soyinka is appalled by ethnic violence and fears the chain reactions which can be unleashed as people define their identities in ever-more-exclusive forms, he is not necessarily troubled by the more general trend in our times toward reaffirmation of the right of self-determination, including secession from existing states. Within an African context, the restructuring of states may in some cases be essential in order to put an end to violence. Soyinka therefore stands both for self-determination and for a renewed effort to construct broader regional bases for solidarity, along the lines of those envisioned by leaders of the struggle for independence from colonial powers. The example of the European Community is of course relevant in this discussion.

IV

In the last analysis, then, the growing concern among many in the liberal democracies of the West that their own states have grown too strong, and must be curbed, stands in sharp contrast to the dilemma of millions of people on the African continent whose states are so weak and inefficient that they can provide neither order nor essential services nor a minimum level of social protection. Nevertheless, there is a common thread of concern in both cases, and that is with defining the balance between individual rights and collective obligations.

Over large parts of the world, authoritarian governments leave little room for individual rights. The 'petro-fundamentalism' examined by Mernissi, the personalistic military regimes of Africa and, until recently, of Latin America,

are all contexts in which elementary attempts to establish a rule of law
provoke repression. Obviously, it is vital to promote respect for the dignity
of the individual in such cases. This is, however, a different endeavour from
that which currently occupies a growing minority within Western liberal
democracies, intent upon asserting almost unlimited rights to live their lives
as they like. In the latter view, virtually any institution in society, from the
family to the school to the local community, can be seen as a 'dominant
structure' or 'oppressive culture' whose underlying values are personally
limiting. The boundaries of interaction with others are reduced to the
minimum necessary to ensure that personal values are not challenged.

Amitai Etzioni's contribution to this volume should be understood against
the background of surging ideological support for extreme forms of
individualism (or cultural exclusivism) in the United States. The 'communitarian' movement, founded by Etzioni, challenges proponents of both
free-market idealism and postmodern insistence upon personal realization to
reconsider the importance of community values. In other words, it attempts
to ensure a balance in that country between what might be called 'libertarianism' and commitment to the common good. This is a central element in
strengthening American democracy — and, in fact, democracy in general —
which depends upon balancing minority rights and majority rule.
Libertarianism is incompatible with the maintenance of a democratic polity.
The 'centrifugal forces' which impel people to pursue their own interests
must be offset by 'centripetal forces' which pull people together in the service
of shared goals.

There are, however, more and less state-centred arrangements for
promoting the common good in modern societies: the arena of the 'public'
includes both governmental and non-governmental institutions, and there
can be many different relations between them. The communitarian movement strongly supports the traditional American pattern of reliance upon
private voluntary activity. Furthermore, Etzioni suggests that, in the future,
all societies 'will need to do more societal business following this model',
because state resources are increasingly insufficient to meet public needs.

It should be stressed that insistence upon strengthening 'civil society'
implies no necessary attack on the state. Although the neoliberal discourse
on development sets up a false dichotomy between 'civil society' and the
state, and then suggests that the former can only thrive if the latter is
restricted, this would apply only to a deeply authoritarian setting. In fact, as
Etzioni points out, a strong civil society is the condition for a strong and
effective democratic state: and in order for interest groups and voluntary
associations of all kinds to interact productively, the existence of an efficient
and legitimate government is essential.

If this balance between liberty and order is difficult to maintain even in the
most established Western democracies, it should not be surprising that it is
even more difficult to construct in societies embarking upon 'transitions to
democracy'. Voluntary private organization has long been repressed in these

cultures and is only now beginning to develop openly. Its future role in providing for the well-being of people and its future relation to new public sector institutions are still uncertain.

In the former Soviet Union, the intelligentsia played the part of an incipient civil society, maintaining a space for freedom of thought and association within a repressive political system. Tatyana Tolstaya tries to capture the essence of this elusive group — a 'multitude of people, from all classes and estates' — who lived in the interstices of the Soviet regime. They were an idealistic moral force, stubborn humanists in a philosophical setting dominated by relentless materialism. Yet, in Tolstaya's view, their very idealism has ill prepared them to cope constructively with the extraordinary challenges of the present period.

The intelligentsia tended to have a romantic view of society, in which 'the people' were 'a mythical, indivisible whole', undistinguished by specific interests and capacities. This myth, maintained by Soviet ideology, did not allow for the comprehension of pluralism and conflict or for the calculation of risks associated with dismantling the former economic and political system. The romantic idealization of foreigners has also served the intelligentsia badly, as their expectations of brotherhood have been shattered by the reality of ethnic conflict at home and *realpolitik* abroad.

Tolstaya presents us with a picture of the intelligentsia in retreat, confused by a chaotic political situation and frightened by the spectres of ethnic conflict and fascism. The former role of dissident must now be transformed, so that those who opted out of the old regime can 'opt into' the new one. In broader terms, all groups within Russian society must learn to speak to each other in a new political language and to propose concrete measures for the reconstruction of their country. The savage atmosphere of primitive capitalist accumulation which currently permeates broad areas of life does nothing to promote a rational construction of common ground.

V

A danger in countries of the former Soviet Union, as in other socialist settings undergoing a process of profound reform, is that the concept of 'freedom' for which Western liberal democracy stands will be confused with the kind of extreme individualism already discussed above. The neoliberal orthodoxy holding sway in parts of the international financial community can exacerbate this danger. In fact, a narrow faith in self-regulating markets, in which the public good is determined by the virtually unrestrained pursuit of private interests, has taken a high toll on public spiritedness and security in most countries where it has assumed a central role in government policy over the past few decades. Dahrendorf describes this development as 'economism run amok' and sees an urgent need to 'change the language of public economics' if further damage is not to be inflicted upon the public sphere.

The underlying assumption in today's radical free-market economics is that leaving people to fend for themselves, without protection or security, is the surest way to promote economic progress. Adequate minimum wages, public programmes of social security and other forms of welfare provision stifle initiative and slow growth. But as Emma Rothschild illustrates in her analysis of late eighteenth-century political economy, this point of view would not be supported — and was not in fact supported — by some of the most outstanding early proponents of *laissez faire*, including Adam Smith.

Smith, Condorcet, Turgot and others were passionate believers in private property, free trade and the limitation of state power. At a time when individual freedom depended upon overturning corporate or monarchical privilege, these were elements in the construction of a more open and just society. Nevertheless, far from linking support for free markets with the kind of Social Darwinism so frequently seen today, Smith and his French followers were also committed to improving the lot of the poor. Unlike current champions of low-wage strategies of development, for example, Adam Smith was a strong supporter of high wages; and he spoke out as well in favour of progressive taxes on luxury goods, so that ' "the indolence and vanity of the rich is made to contribute in a very easy manner to the relief of the poor" '.

Rejecting the kind of private charity which demeaned those in need, Condorcet went on to design the first schemes of public social insurance, in order to work toward creating a different sort of society, ' "something which has never before existed anywhere, a rich, active, populous nation, without the existence of a poor and corrupted class" '. Freedom was associated with justice — a moral concept which the great figures of early *laissez faire* economics were not afraid to discuss.

Taking up this eighteenth-century debate on political economy, many of the participants in the Copenhagen conference argued forcefully for restoring the discipline of economics to its full standing as a social philosophy, in which promoting people's sense of security, well-being and self-respect has always been as important a subject of concern as generating economic growth. As Emma Rothschild insisted, there is no compelling reason why even free-market economics must be associated with 'a flint-hearted view of society, in which men and women are surrounded only by incentives, and inspired only by fear'.

In his intervention at the conference, Tetsuo Najita proposed exploring the relation between economics and morality by going beyond the 'cosmopolitan discourse' of Western intellectuals, to focus upon 'the thought and practices of ordinary people in the everyday world of ethics, contract and commerce'. Like Rothschild, he takes a historical example, again from the eighteenth century. But the country this time is Japan and the subject is the network of village credit and insurance co-operatives that came to have an enduring role in Japanese social history.

These organizations grew up in a context of periodic famine and natural disaster not dissimilar from that experienced by commoners in Europe during the eighteenth century. Since the feudal regime in Japan could not be relied upon to deal with these threats to security, people came together and pooled their resources for the purpose of 'saving each other' in times of emergency. Such institutions were sustained by moral contracts with no legally binding status; they had nothing to do with the state and did not respond to market signals. Yet by the early 1900s, there were almost 334,000 local contract co-operatives throughout the country, serving as a crucial credit infrastructure for ordinary citizens under conditions of capital scarcity. Najita uses the example of the Japanese village co-operatives (and their Danish counterparts, much admired by Japanese visitors) to illustrate the enormous creative potential of social solidarity, where equality — not the lack of it — drives economic progress. The philosophy underlying membership in contract co-operatives was not profit maximization, but a reverence for life as a gift of the natural order. Thus ethics and economics were intertwined, as in fact they ordinarily are when one descends from theory to the daily life of people, whether traditional or modern, Northern or Southern, Eastern or Western.

VI

Ethics assumes particular importance when, as Anthony Giddens explains in his essay, the rapid advance of science and technology creates dilemmas for which no strictly technical solution can be found. Value judgements are then particularly salient in determining choices. This is the case today, as our increasing capacity to shape the physical world requires us to make decisions on issues which in the past would never have arisen (many in the field of genetic engineering provide a particularly useful example) or in areas which, given our earlier limited capabilities, would simply have been left to nature or chance. The need to consider alternatives and take decisions is also growing within the sphere of interpersonal relations, as basic social institutions (like marriage and the family) evolve rapidly and traditional mores are questioned.

The very complexity of the decision-making field in which modern human beings live — the elimination of fixed points of 'destiny' and the 'natural order' — contributes to the sense of disorientation that marks our times. There is also a heightened sense of risk, associated with a technological revolution which seems to spiral out of control and to threaten the natural balance of the planet. Giddens calls this 'manufactured risk', attributable to human invention, to distinguish it from 'external risk' or chance.

In this context, politics is taking on new dimensions. In addition to the long-established practice of 'emancipatory' politics (centred around the promotion of social justice through the redistribution of scarce resources), a new concern with 'life politics' emerges. The latter is not primarily redistributive; it involves creating coalitions to deal with modern, 'manufactured' risk and to improve the quality of life. But although shifting resources from rich to poor is not its central purpose, Giddens suggests that 'life politics' can have emancipatory implications. Political mobilization on life-style issues has the capacity to build common ground among broad categories of people, not only within countries but across national borders. For example, concern with environmental, gender and some consumer issues cuts across classes and nationalities. Rich as well as poor have an interest in improving the quality of the air, removing harmful pesticides from foods, improving the life chances of women. As Giddens notes, 'few things can be more significant worldwide than the possibility of a new social contract between women and men, since sexual divisions affect so many other forms of stratification in societies of all types'.

The growing importance of 'life politics' leads Giddens to speculate about the possible appearance of what, in his view, might be called a 'post-scarcity society'. This would not be a society without want, but one in which expanding awareness of the dangers of technological change, and increasing understanding of our common problems (including problems of self-definition), promote attempts to improve people's lives outside the bounds of confrontation over redistributive issues. For example, many institutions in society can be strengthened, or reconstructed, without necessarily making use of greater economic resources. People who have attained a certain minimum standard of living can enjoy healthier and more satisfying lives by taking measures to change their life style which cost nothing.

It is also possible to create coalitions in which something less preferred by one party is traded for something more preferred by another: a kind of life-style barter, in which every participant gains. Such is the case, for example, when the old and young in atomized and lonely neighbourhoods come together to offer the experience and loving attention of the former in exchange for the physical assistance of the latter. They are not engaged in a zero-sum game, nor are transactions defined primarily within the framework of 'productivist' economic competition. Thus undertakings of this kind can become elements in the construction of a 'post-scarcity society'. Such a proposal forcefully reminds us that societies are changed by very personal decisions about how we want to live. It evokes the countless examples of co-operation (both privately- and publicly-funded) which improve the quality of life of people through exploiting synergies. It emphasizes the central role played by international non-governmental organizations, working on environmental and social issues, in influencing contemporary patterns of societal change.

This is a moderately optimistic picture. In contrast, Johan Galtung's vision of the future, presented in the final essay, is the most pessimistic of any in the collection. In his view, our present course sets humanity 'on its way from nomadism to monadism' — a structureless and cultureless society in which individuals, thinly and weakly related, act out of egocentric cost-benefit analysis. The large, bureaucratic structures of modern societies are becoming unworkable — gutted by too great a reliance on cybernetics — and the small, personal domains in which people can feel needed and loved are being torn apart. This can only promote growing violence, corruption and mental disorder. There is something of Anthony Giddens' erratic 'runaway' world in this formulation, but Galtung carries the metaphor to its theoretical extreme of chaos and despair.

Nevertheless, Galtung explores constructive ways to adapt to forces of detraditionalization and globalization. In particular, he stresses the importance of 'rehumanizing' the vast impersonal institutions of modern society, including transnational corporations and government agencies, and strengthening the smallest units of long-term, personal interaction or 'belonging', however these may be composed. He also turns his attention to religion and stresses the need to reinforce 'binding normative culture' in a world seriously threatened by anomie. For Galtung, the traditional disdain shown by secular humanism for religion is misplaced. Religion plays a fundamental role in providing a meaning for life; and if human beings are to find a way out of their current predicament, they must take every opportunity to strengthen and support a universal 'culture of compassion'. Since all world religions contain both an aggressive, exclusive component and a less aggressive, inclusive one, 'the most important struggle in the religious landscape of the world' today is not among religions at all, but within each of them. Reinforcing the tolerant, compassionate currents within each religious community is one of the most important tasks of our times.

VII

Where is the world heading, then, at the turn of the twenty-first century? Toward a desolate landscape of disintegrating societies, where urban gangs pillage great Western cities and marauding military bands roam Third World countrysides? Toward spiralling international violence, based not (necessarily) upon Cold War enmities but upon escalating ethnic conflict and harsh economic competition? Or toward construction of new institutions to mitigate violence, reduce poverty and promote solidarity in a rapidly changing global context?

The way one approaches such questions depends, of course, on ideological commitments and on the intellectual boundaries associated with particular scientific paradigms. None of the contributors to this volume would accept the orthodox free-market projection of future trends, in which increasingly unfettered markets — reflecting increasingly unlimited individual choice — would shape a new world of democracy and plenty. This neoliberal view flies in the face of evidence from around the world that radical free-market restructuring increases inequality, deepens poverty and heightens conflict. Democracy has never flourished without strong institutions to guarantee certain minimum standards of socio-economic and political equality. Nor would most participants in the Copenhagen conference subscribe to the defeatist attitude of postmodernism, holding that since it is clearly impossible to understand the human condition in all its complexity, there is no point in trying to develop any hypotheses on social change at all. This perspective leaves us with a chaotic world in which any proposals for improving economic and political institutions are impossible.

The picture emerging in this volume is a predictably complicated one of shifting patterns of social integration and disintegration, creating enormous challenges for human understanding and ingenuity. Basic institutions of society, from the nation state to the family, are under stress. The boundaries of solidarity are being redrawn, within communities and across the world. People are redefining their values in a period of fundamental ideological and religious questioning. Some important elements of this crisis are global; but one of the points made forcefully in the essays which follow is the need to understand local situations and to design local strategies for dealing with particular problems. 'The state' — to take one example — has specific characteristics in concrete settings; it may therefore need to be strengthened in some cases (and sectors) and weakened in others. The balance between individual rights and the common good sometimes requires the provision of greater support to the first element in that equation, and sometimes to the second. The tolerant elements in all religions require support, while the intolerant aspects of the same religious institutions do not. In other words, many attempts to offer universal guidelines for creating a better world are at best irrelevant, and at worst dangerous, when applied indiscriminately to very different cultures and societies. For this reason, a dialogue within and across cultures is essential.

Participants in the Copenhagen conference have highlighted a number of issues deserving wide debate. They warn us to take seriously the threat of authoritarianism, and even of fascism, at the end of the twentieth century. They suggest that we explore the relation between global economic competition and a tendency to shore up regimes that can ensure certainty in extremely uncertain times. They note new forms of global dependence, in which essential decisions affecting the livelihood of millions are taken by international financial institutions and transnational firms only minimally accountable to any broad political constituency.

The essays in this volume also suggest the urgent need to reconsider the meaning of work in modern society and to design new ways to provide for minimum levels of well-being which do not depend entirely on remunerated employment. Authors call for a 'positive critique' of the welfare state and for the development of national strategies to halt the process which tends to exclude ever-larger numbers of people from productive participation in the economy. They recognize as well that in a highly interdependent world system, international initiatives to protect the livelihood of all people must be explored.

These strategies require negotiation within and among nation states; in other words, they must be carried out within relatively open and efficient political systems. The authors of these essays are also concerned, however, with exploring the capacity of people to develop new structures of mutual support and exchange within a non-governmental context. They provide numerous examples of how the institutions of 'civil society' can and do provide for the common good; and they discuss the potential for creating new alliances along class and non-class lines within the international community. There is, in the following pages, a repeated reference to universal human values, perhaps most simply and directly summarized in the moral contract drawn up by eighteenth-century Japanese villagers and transcribed by Najita: 'Think of the misfortunes that happen to others around you as though they were your own'. Treat others as you would like to be treated. These values are among the most powerful moving forces in the history of mankind.

There is nothing in the current global context which cannot be improved through concerted action. The particular form of 'globalization' currently shaping our lives — with its overriding emphasis on competition and its degrading lack of concern with human security — is not immutable. It is the product of adherence to an ideology that interprets life as a vicious struggle to be won by the strongest. Such a world view requires modification. Human beings are motivated by solidarity and hope, as well as by selfishness and fear. This is not a new insight; but it bears reaffirming. If taken more fully into account, it can alter the parameters of social development in the twenty-first century.

Economic Opportunity, Civil Society and Political Liberty

Ralf Dahrendorf

IN DEFENCE OF THE FIRST WORLD

At its best, the First World was not a bad place in which to live and to thrive. Did anyone ever call it the First World? Or was the numeral merely the backdrop for the unmentionable Second World of communist oppression which has now all but disappeared, and the Third (later also the Fourth) World of destitution, disease and despondency? Whatever the motive, let us not dismiss the First World too easily. At its best, it combined three social virtues:

- economies which not only offered a decent life to many but which were set to grow and to open up opportunities to those not yet prosperous;
- societies which had taken the step from status to contract, from un-questioned dependence to questioning individualism, without destroy-ing the communities in which people lived;
- polities which combined respect for the rule of law with those chances of political participation, of dismissing as well as choosing governments, which we have come to call democracy.

One may well ask when and where such wealthy, civilized and enlightened countries existed. The temptation is considerable to hide behind acronyms and refer to what is often called the OECD world nowadays, the membership of the Organization for Economic Co-operation and Development. But let me resist the temptation and name names. The United States of America in the period from Roosevelt to Kennedy, if not to quite the same extent before and after, is one example. Tens of millions of people from all over the world dreamed of living in America, and millions went to great lengths to get there. Magnets for immigration are not the worst index of social well-being. This applies to other countries as well. The United Kingdom has long had a more even balance of migration — except for the Irish, for persecuted Jews and later for people from the poorer colonies — than the United States; but for long periods of this century it certainly belonged in the First World as here defined. So did parts of the former British Empire, the "temperate Common-wealth" as some call it in geographically correct if politically incorrect language — Australia and New Zealand, Canada and a few other bits and pieces around the world. Then there are smaller European countries to mention: Switzerland, Sweden, and the other Scandinavian states. By the 1950s, when the Organization for European Economic Co-operation (which meant, above all, reconstruction) was turned into the OECD, most of Western Europe had become a part of the 'happy few'.

Their characteristics were, to repeat, economic opportunity, civil society and political liberty. However, it would be testing the benevolence of the reader beyond the permissible to leave such smug statements without qualification. In fact, three major qualifications have to be added before a serious discourse becomes possible. Each of these qualifications would warrant an essay of its own.

First of all, the perfection of the First World in its heyday was flawed. All of its members excluded some from the benefits of their achievements, and even from opportunities. The history of the United States is one long sequence of battles for inclusion — from the Civil War to the Civil Rights campaigns and beyond, to today's underclass. For the most part, the battle could be fought within the institutions of the country, which is worth noting. Moreover, it was fought not just by the excluded themselves; they had allies, in the Supreme Court for example, which is also worth noting. But American society was never even nearly perfect in terms of economic opportunity, social inclusion, or political participation. To the present day (to mention just one of many shocking facts) the American president is probably elected by no more than 15 per cent or so of those who, by law, should be entitled to vote.

The American imperfections are stark and visible, but those of the United Kingdom or Australia, Switzerland or Sweden are no less important. Economic inequality meant for many that the promise of citizenship remained unreal. The social conflicts which would presumably have domin-ated a world summit on social development a hundred years ago were fierce; government representatives at an 1895 summit would for the most part have recommended the suppression of the conference by force. It took decades of internal struggles — class struggles as they were correctly called at the time — to assert the basic equality of all human beings in society. It also took two modern wars because, horrible though it is to say this, there is no greater social equalizer than a modern war in which entire populations get involved. It was not an accident that the Second World War was called a 'total war'.

These wars, to be sure, were not fought by the great democracies among themselves. They set civilized and not (not yet?) quite civilized countries against each other, those which had made it in terms of turning opportunities into general entitlements, and those which had not quite made it. I stress this point advisedly, and will even add a general thesis: the greatest risk to peace emanates from countries on the way from the old cycle of poverty, dependence and illiberty to the life chances here described as those of the First World. When opportunities are held out for people but are not yet there to grasp, when economic development accelerates but social and political development lags behind, a mixture of frustration and irresponsibility develops which breeds violence. Such violence can be individual and undirected, but it can also become collective and directed against apparently happier neighbours, or more successful strangers in one's midst, or both. While it is likely that economic development coupled with political democracy and a civil society generates both an internal sense of tolerance and peaceful international relations, the road which leads to such a state is full of pitfalls and temptations. *Imperial Germany and the Industrial Revolution* (to quote Thorstein Veblen's 1915 title) is only one example. Whenever a formerly traditional country embarks on this road, the rest have good reason to be apprehensive as well as hopeful.

Yet this is not said to condemn the rest to poverty: on the contrary. The second qualification of my initial thesis about the First World is that civil society — citizenship — is incompatible with privilege. This holds not just at home — in a given country, where privilege is by the same token a denial of the citizenship of others — but internationally too. As long as some are poor, and moreover are condemned to remain poor because they live outside the world market altogether, prosperity remains an unjust advantage. As long as some have no rights of social and political participation, the rights of the few cannot be described as legitimate. Systematic inequality — as against comparatively incidental inequality within the same universe of opportunity — is incompatible with the civilized assumptions of the First World.

This is a moral statement, but it is not just a moral statement. Take immigration, which tells the whole story. To obstruct the free movement of people is in principle unacceptable for free countries. Yet one appreciates that Switzerland, for example, would put the quality of life of its citizens at risk if it allowed everyone who wanted to come actually to settle on equal terms. So what does the country do? It allows some in because they can make a useful contribution by, for example, enabling the locals to shirk disagreeable jobs; but it turns them into second-class citizens who cannot vote and who can be sent 'home' at short notice. Most, however, are not allowed in at all; and to implement such a policy, a whole machinery of control has to be set up not only at the borders but also within the country. The humiliating experiences of asylum-seekers in many countries of the First World are an indictment of the latters' claims to civilization, and yet there is no simple answer to the predicament.

Rather, there is only one answer, and it is not simple. It is the universalization of the benefits of the First World — what has come to be called development. Others are more qualified than I am to express a view on this vast set of issues. We now know for sure, if we did not know before, that economic and social development is as much a matter of internal effort as it is of external assistance. We also know that large countries, notably in Asia but also in Latin America, have embarked so successfully on the road of economic development that the old First World is beginning to regard them as a threat. When people speak of the Third World these days, they mean, very largely, Africa; and by Africa they do not mean Tunisia or the liberated South Africa. Thus development can, and does, happen.

However, it is not only a precarious, but also a long process. Arguably, it takes mankind through the most threatening period of its history. The so-called population explosion; the dangers of military aggression, aggravated by the wide diffusion of lethal, even nuclear, weapons; militant *integrisme*, the French term which is preferable to 'fundamentalism' because it emphasizes the non-differentiation of religion and secular concerns such as the rule of law; protectionism with regard to goods as well as to people — these and other evils are all possible, and all too often real by-products of the early phases of development, and they will be with us for generations to come. Yet the process is necessary, not because of any hidden hand of History (such Hegelianism is far from my Popperian way of thinking) but because the very values of an enlightened and civilized society demand that privilege be replaced by generalized entitlements, if not ultimately by world citizenship then by citizenship rights for all human beings in the world.

Add to these two qualifications a third one, and the seemingly bright picture with which I began looks more overcast still. Actually, the third qualification has a great deal to do with Karl Popper, or with Herakleitos long before him: παντα ρει — everything is in flux, nothing lasts, not even the blessings of prosperity, civil society and democracy. It was not without reason that I used the past tense in referring to the achievements of the United States, the United Kingdom, and even Switzerland and Sweden. At times one has a sense that the great age is over; at any rate, it is under threat. Having set the wider scene, it is these threats to the First World on which I want to concentrate, before turning to a few modest recommendations for combating their effects, and perhaps their causes.

Put at its crudest, the countries of the OECD world have reached a point at which the economic opportunities of their citizens lead to perverse choices.

In order to remain competitive in growing world markets, they have to take measures which damage the cohesion of civil societies beyond repair. If they are unprepared to take such measures, they have to resort to restrictions on civil liberties and political participation which amount to a new authoritarianism, no less. At least, this appears to be the quandary. The overriding task of the First World in the decade ahead is to square the circle of wealth creation, social cohesion and political freedom. While completely squaring the circle is impossible, one can get close to it — which is probably all a realistic project for social well-being can hope to achieve.

Perhaps countries outside the charmed circle will find a way through this maze first, though it is hard to think of many. Mexico and other success stories in Latin America? The probability must be high that they will share the European-American malaise. The countries of the former Second World which are now post-communist? They are clearly struggling with every one of the three tasks — economic opportunity, civil society and political liberty. The tigers and dragons of Asia? China? For the time being, almost all these countries reject the problem as phrased here in that they seek rapid economic growth combined with strong social cohesion without trying unduly hard to advance the rule of law and political democracy at the same time. Thus we are back to the OECD world — including its members far and wide, such as Japan — if we accept the project.

GLOBALIZATION, ITS CONSTRAINTS AND ITS CHOICES

It would be possible to consider economy, society and polity separately; in fact this has often been done. Economic growth is uppermost on government agendas in the OECD world, and their advisers — civil servants as well as professors — help them focus on this issue to the exclusion of all others. Will deregulation do the trick? Is inflation, after all, a helpful lubricant? How do taxes have to be levied in order to stimulate rather than to impede growth? Extreme proponents of 'economism' — economics as a political ideology — not only ignore but decry social factors. Was it not a prime minister who said that 'there is no such thing as society' because she wanted to encourage individuals to fend for themselves? But society has enough advocates even at a time when sociologists are viewed with a suspicion sharpened by 1968 and all that. In any case, the dissolving, disintegrating quality of modern societies has been a theme for a century. Anomie, suicide, crime; the collapse of the family; the loss of religion — these were themes long before 'community' became a correct word again. And so far as the polity is concerned, democracy has been in crisis as long as anyone can remember such a thing as political science. Governability was certainly an issue in the 1970s; but long before that scholars, and politicians, were wondering why people turned against democracy when unemployment hit them and the stock market collapsed.

Such allusions may help avoid the fallacy of historical uniqueness — though the opposite fallacy that history keeps on repeating itself is no less risky. There is a case for saying that in the OECD world, economic, social and political well-being are intertwined in a new and vexing manner. The reason is probably, in one word, globalization. It has become hard, and for most impossible, to hide in this world. All economies are interrelated in one competitive market-place, and everywhere the entire economy is engaged in the cruel games played on that stage. There is literally no getting away from it, and the effect of globalization is felt in all areas of social life.

The sceptic will no doubt raise his or her eyebrows: is this really so? And why should it be so? Moreover, what exactly does globalization mean? The sceptic would win, as of 1995, a good part of the argument. Globalization is, so far, by no means total. Whole economies, including that of China, are more national than global (though part of their national success is due to their global involvement). Economic regions are forming to provide common markets or free trade areas (although this may be a response to the new productive forces of globalization rather than their refutation). Within countries, important activities, like the provision of health services or of nursery and primary education if not education in general, seem removed from global competition (though it cannot be an accident, or a mere fad, that the values of a globalizing economy have entered these services). It would certainly be possible to make a case for the imperfections of globalization at this point, though whether this would also be a case against its force is another matter.

Why should globalization have happened at all, and why now? The obvious answers are probably the best ones. Whether the end of the Cold War is cause or effect may be a moot question; certainly the Soviet bloc countries were economically no longer viable. One reason was that the concept of country, or nation, lost a good part of its economic meaning. This in turn was a result of the emergence of transnational entities which found it surprisingly easy to combine a degree of adjustment to local needs with the advancement of worldwide strategic planning, direction, and profit-making. Add to this the two (related) 'revolutions' of information technology and financial markets, and an economic scene emerges the likes of which the world has never seen before. Not only in terms of movements of money, but also in terms of services (like booking airline tickets), and in the end even in terms of production, conventional physical boundaries begin to lose all meaning. Politics and technology, market pressures and organizational innovations all conspire to create, in important areas of economic activity, a wholly new space which anyone — any company, any nation — ignores at their peril.

What then does such globalization mean? The most important answer is to the question not asked: what does it **not** mean? There has never been — as Michel Albert reminded us in his *Capitalism vs Capitalism* (1993) — just one economic culture, even among the market economies. We have long sensed that Japan is different from America, and Germany from the United Kingdom. The differences are quite profound, even if they are badly understood. The rest of the world keeps on pressing Japan to open up its markets, when one cause of their inaccessibility is people's ingrained tastes and another is the Japanese language. Germany and the United Kingdom speak much the same language of economic policy, yet the textbook capitalism of the United Kingdom and the textbook corporatism of Germany create very different attitudes, especially since in the first case the textbook is one of economics and, in the second, one of political science.

Such cultural differences will not disappear. To what extent they will in future be national, may be an open question. There is a certain cultural plausibility, after all, to the regions which are beginning to form, those of Europe, of the Americas, and of East and South-East Asia. (It is also apparent that certain countries do not obviously belong, such as the United Kingdom in Europe and, perhaps increasingly, Japan in Asia.) Whatever the shape of the structures which finally emerge from the moving kaleidoscope of not-yet-crystallized world affairs, the basic presumption remains that reactions to globalization will differ despite the fact that the global market-place requires some of the same virtues from all. Indeed, if it were not for such differences, the question raised in this paper would lose its meaning. Squaring the circle of economic growth, civil society and political liberty is a universal task, but it would be foolhardy to assume that everybody will tackle, or even try to tackle it in these terms. For those who will, the assumption is that the goal can be approximated without losing out in the global market-place.

What then are the inescapable conditions of globalization? What, in other words, has to be done everywhere if companies, countries or regions do not want to condemn themselves to backwardness and destitution? To use the fashionable word, economic actors need above all flexibility. The word is intended to convey something desirable, though for many it describes the price they have to pay. Also, the word has so many connotations that it is hard to pin down to any particular meaning. Yet without considerable flexibility, companies cannot survive in the world market.

Flexibility means in the first instance the removal of rigidities. Deregulation and less government interference generally help create flexibility; many would add a lighter burden of taxation on companies and individuals. Flexibility has increasingly come to signify the loosening of the constraints of the labour market. Hiring and firing become easier; wages can move downwards as well as upwards; there is more and more part-time and limited-term employment; workers must be expected to change jobs, change employers, change locations of employment. They have to be flexible themselves. So do entrepreneurs of course; Schumpeter's idealized figure of the entrepreneur and his 'creative destructiveness' is invoked. Flexibility also means the readiness of all to accept technological changes and respond to them quickly. In marketing terms, flexibility is the ability to move in wherever an opportunity offers itself, and also to move out when past opportunities close. The story is familiar enough, as is its accompanying language of structural adjustment, efficiency gains, competitiveness and seemingly unending increases in productivity.

Yet choices remain. At least they are choices in theory; in practice they are just as likely to be brought about by circumstances, traditions and irresistible pressures. Let me mention two such choices because they are relevant for the argument of this paper. One is that between a low-pay and a high-skill economy. In practice, most countries will combine both in some ways, but there are important differences of emphasis. Low-pay economies find their place on the world market by undercutting others. Their products are cheaper, though their workers are also poorer. One can sometimes hear arguments that this is the only road to success; but the evidence is that such extreme economism is simply a mistake. High skills can also create a competitive advantage. This is not just the case because by high skills, and only by high skills, the frontiers of technology are advanced, but also because despite computerization certain quality products and product qualities require a skill input. Indeed there even comes a point at which one highly skilled person is cheaper than five low-paid ones who produce the same effect. In terms of such choices, the United States seems to move in the direction of low pay, whereas Japan opts for high skill; the United Kingdom prefers low pay and Germany high skill.

Another choice may well be related to the first; it is even harder to describe precisely. This is the choice between low taxes and other contributions accompanied by high yields on the one hand, and high taxes and contributions coupled with low distributed yields on the other. Crucially, the difference can be investment-neutral in the sense that the funds in question are distributed differently between, say, shareholders and workers, with investment held constant. In practice, the signs are that the low-profit route makes long-term investment if anything more probable than the low-taxation route. However, modes of investment will vary; the low-profit route is likely to involve a greater role for banks and a lesser one for the stock market, and vice versa. Other implications of the alternative would be worth pursuing, such as those for the private or public nature of welfare provisions. In any case, an important choice can be made; and again we see, in practice, the Anglo-American economies (and those which have followed their lead, or perhaps merely the textbooks based on their experience) on one side, that of low taxes and high distributed profits, and Japan and Continental Europe on the other.

One other point needs to be reiterated here, a point made in connection with incomplete globalization. Despite the enormous force of the global market-place, there is and always will be such a thing as privileged access to markets. Even without explicit protectionism, the Japan phenomenon exists everywhere to a greater or lesser extent. People will 'buy American' in the United States, 'buy German' in Germany, and even 'buy British' in the

United Kingdom if there are British products on the market. Regional trade arrangements are about extending these privileged markets. 'Buy American' then means 'buy NAFTA-American', and 'buy European' replaces national rallying calls. In ideal typical terms, such regionalism introduces inflexibility; people will probably call it predictability, or security. The point should be borne in mind.

Speaking of choices must not, however, detract from the main point at issue. The options alluded to here are minor variations on the major theme of globalization. The forces of globalization are strong everywhere. They bring with them pressure for greater flexibility, with all the implications listed earlier. By choosing one or the other variant, companies — even countries, for many of the choices invite government action — can take the edge off certain effects or give additional edge to others; but one thing they cannot do is opt out of the global market-place. Even the attempt to stay in an older socio-economic age to serve the political purposes of dictators will not work for any length of time, as shown by the examples of Myanmar or Cuba, and probably soon the Democratic People's Republic of Korea.

CIVIL SOCIETY UNDER PRESSURE

The next question to address is: what does globalization do to civil society? The answer is that it threatens civil society in a variety of consequential ways. The term 'civil society' is more suggestive than precise. It suggests for example that people behave towards each other in a civilized manner; the suggestion is fully intended. It also suggests that its members enjoy the status of citizens, which again is intended. However, the core meaning of the concept is quite precise. Civil society describes the associations in which we conduct our lives, and which owe their existence to our needs and initiatives rather than to the state. Some of these associations are highly deliberate and sometimes short-lived, like sports clubs or political parties. Others are founded in history and have a very long life, like churches or universities. Still others are the places in which we work and live — enterprises, local communities. The family is an element of civil society. The criss-crossing network of such associations — their creative chaos as one might be tempted to say — makes up the reality of civil society. It is a precious reality, far from universal, itself the result of a long civilizing process; yet it is often threatened, by authoritarian rulers or by the forces of globalization.

The social effects of economic responses to the challenges of globalization have become the subject of public and scholarly attention, especially in the United States. This is no accident. North America is the home of modern civil society, where threats to its strength are most acutely felt. Suddenly, Tocqueville's world, indeed that of the Federalist authors, appears to crumble. *The Disuniting of America* is the new theme (Schlesinger, 1992), accompanied by fear, violence and versions of fundamentalism. It is little consolation that America is not alone in this predicament. The following sketchy catalogue of pressures on civil society draws from European as much as American experience, and is at least in part applicable to other OECD countries as well.

Economic globalization (to begin, without any particular reason, at one relevant point of the story) appears to be associated with new kinds of social exclusion. For one thing, income inequalities have grown. Some regard all inequalities as incompatible with a decent civil society; this is not my view. Inequality can be a source of hope and progress in an environment which is sufficiently open to enable people to make good and improve their life chances by their own efforts. The new inequality, however, is of a different kind; it would be better described as inequalization, the opposite of levelling, building paths to the top for some and digging holes for others, creating cleavages, splitting. The income of the top 10, or even 20 per cent is rising significantly, whereas the bottom 20, indeed 40 per cent see their earnings decline. Robert Reich and others have made this observation the starting point of their search for remedies though even the US president's labour secretary has not been able to do much to reverse the trend. The systematic divergence of the life chances of large social groups is incompatible with a civil society.

The process is aggravated by the fact that a smaller but significant set seem to have fallen through the net of citizenship altogether. The concept and the phenomenon of the underclass is much discussed. Not everybody likes the term, which is clearly misleading if one considers it in terms of class theory. The socially excluded are not a class; they are at most a category of people who have many different life stories. Though some of them manage to get out of the predicament, many are in a position in which they have lost touch with the 'official' world, with the labour market, the political community, the wider society. Do they number 5 per cent, 10 per cent? Figures vary, but most OECD countries now have in their midst what William Julius Wilson (1987) called 'the truly disadvantaged' — would-be citizens who are non-citizens, an indictment of the rest.

Many of the truly disadvantaged are not just economically excluded; they are also excluded on other grounds, as 'strangers' by virtue of race, nationality, religion, or whatever distinguishing marks are chosen to provide excuses for discrimination, xenophobia, often violence. Declining social groups, like the 40 per cent whose real incomes have been falling for ten years or more, are the breeding ground for such sentiments. Borders, including social boundaries, are always particularly noticeable for those closest to them. A wave of 'ethnic cleansing' is not confined to war zones like Bosnia and Herzegovina, but threatens to engulf us all.

What does this have to do with globalization? So far as the new inequality — the increasing divergence of those near the top and those near the bottom — is concerned, it takes us back to the low pay-high skill option. Those whose skills are needed are paid a good salary, but many who had a reasonable wage or salary in the past have now sunk to a miserable and often irregular real income. Nor is there any evident route back to the upward slope. Hard as this may sound, some are simply not needed. The economy can grow without their contribution. Whichever way you look at them, they are a cost to the rest, not a benefit.

Milder — although, if anything, more acutely felt — versions of the experience have now hit the middle classes. The latest wave of efficiency gains has meant, especially for large companies, making office workers redundant, all the way to the once hailed echelons of middle management. Such trends document a fundamental change in the world of work. No one would argue that there is not enough work to be done, but work at decent rates of pay is increasingly hard to come by. It is a privilege, not a realistic aspiration for all. Manufacturing and many services are following agriculture into a stratosphere of productivity in which half or fewer of those employed in the past can produce twice as much output or more. What remains is a strange assortment of ill-paid personal-service jobs, numerous forms of hidden unemployment — some called 'education', others 'self-employment' — and, in Europe, long-term unemployment for at least 5, and probably before long 10 per cent of the population of employment age.

I have not offered here any more probing thoughts about the conditions under which civil societies thrive. However, poverty and unemployment threaten their very fabric. Civil society requires opportunities of participation which in the OECD societies (if not universally) are provided by work and a decent minimum standard of living. Once these are lost by a growing number, civil society goes with them.

Let me move on to another set of social issues associated with economic globalization. The ambivalence of flexibility has been mentioned. It may be the other side of rigidity, but it is the reverse of stability and security as well. One may fairly debate the extent to which stability and security are pre-conditions of civil society. Both geographical immobility and welfare state security may have gone too far in parts of Europe in the 1960s and 1970s. The American experience of the past shows that effective communities can be created despite high mobility and low welfare provision. But the economic response to globalization is intrinsically inimical to both stability and security. Uprooting people becomes a condition of efficiency and competitiveness; the 'get on your bike and look for work' mentality rules. In addition, the dismantling of the welfare state is on the agenda everywhere.

Such developments may not be all bad; they are to some extent unavoidable. But the pendulum is swinging far in the opposite direction. The dual effect is the destruction of important features of community life and a growing sense of personal insecurity for many. Inner cities tell a shocking part of the story, aggravated by the tendency to erect green field shopping centres at the expense of high streets and market squares. Limited-term contracts — like part-time work — are fine for a while, notably for the young and the able-bodied and perhaps for child-bearing women; but people, even children, do get older, and discovering at the age of fifty-five and sometimes earlier that you are no longer needed is enough to turn many into 'grey panthers'.

Add to such phenomena the return of Social Darwinism under the pressures of globalization, and the concoction becomes even more lethal. At times one detects strange similarities, at least in Europe, between the end of the nineteenth and the end of the twentieth centuries. Then as now, people had been through a period of rampant individualism — Manchesterism then, Thatcherism now. Individuals were set against each other in fierce competition and the strongest prevailed, or rather those who prevailed were described as the strongest, whatever qualities had led them to their success. Then, as now, there was a reaction. Around 1900, it was called collectivism. Today, this is a badly discredited word. However, the new vogue has a similar objective; it is called communitarianism.

Perhaps the most serious effect of the values which go with flexibility, efficiency, productivity, with competitiveness and profitability, is the destruction of public services. The term should be disentangled: the destruction of public spaces and the decline of the service values that go with them. The prevailing carrot-and-stick philosophy has overlooked and then attacked those other motives which lead people to do things because they are right, or even because people have a sense of duty, a commitment. Introducing pseudo-economic motives and terms into public spaces robs these of their essential quality. A national health service, universal public education, basic income guarantees under whatever name become victims of an economism which is running amok. Small wonder that commuter transport, or environmental protection, or public safety suffer in the process.

This gloomy picture is not the whole story, of course. Many people are better off than ever before; they have more choices not just of dishwasher fluids and television channels but of education and leisure pursuits; they live longer; they grumble but then people would, and perhaps they should if it helps them to do something about the objectionable things they see. (Readers of Albert Hirschman [1970] will have noticed more than a trace of exit, voice and loyalty in these comments.) Yet there can be little doubt that the economic challenges of the global market-place have not helped civil society. One analytical footnote may help put flesh on this assertion and also provide a link to the issues of political liberty to which we must turn.

I have spoken of the underclass, of grey panthers, and of people giving voice to their concerns. Why is there no massive movement to defend civil society? Where is the twentieth century equivalent of the labour movement of the late nineteenth century? It does not, and it will not, exist. For reasons which antedate the challenges of globalization, individualization has not just transformed civil society, but social conflicts too. Many people may suffer the same fate, but there is no unified and unifying explanation of their suffering, no enemy which can be fought and forced to give way. More importantly, and worse still, the truly disadvantaged and those who fear to slide into their condition do not represent a new productive force, nor even a force to be reckoned with at present. The rich can get richer without them; governments can even get re-elected without their votes; and GNP can rise and rise and rise.

Individualized conflict is by no means easier to handle — to regulate — than organized class or other struggles; on the contrary. It means that people have no sense of belonging, no sense of commitment, and therefore no reason to observe the law or the values which support such a sense. If there are no jobs, why not smoke pot, go to rave parties, steal cars to go on joy rides, mug old women, beat up rival gangs and, if need be, kill. The term 'law and order' covers a multitude of sins, and it is not always easy to give it a factual basis. But it would seem hard to dispute the observation that social disintegration has become associated with a degree of active disorder. Young men, increasingly young women too, and many who are not so young see no reason to abide by allegedly prevailing rules which for them are the rules of others. They opt out of a society which has pushed them to the margin already. They become a threat. Those who can afford it, pay for their protection. No profession is growing faster than private security services (though it is ill-paid and therefore full of temptations). Those who cannot afford protection become victims. A sense that something has gone badly wrong is spreading, a sense of anomie (Durkheimians might say) or lawless and deep insecurity.

TEMPTATIONS OF AUTHORITARIANISM

The condition of global competitiveness coupled with social disintegration is not favourable to the constitution of liberty. Freedom and confidence go well together — confidence in oneself, in the opportunities offered by one's environment, and in the ability of the community in which one lives to guarantee certain basic rules, the rule of law. When such confidence begins to crumble, freedom soon turns into a more primordial condition, the war of all against all. Who thrives in a state of anarchy? The warlord, the impostor, the speculator, the jester if he is lucky enough to find a protector — but not the citizen, for he no longer exists. Everyone else becomes a victim. People do not like the prospect, especially if they had once been citizens. They begin to doubt the wisdom of the fathers of their constitutions if liberty leads to anomie. They look for a way out, for authority.

Again, it is important to turn dramatic and metaphorical language into precise analysis. One obvious aspect of globalization is that the OECD countries are no longer alone in the world. Competitiveness no longer means that Europe and North America have to keep up with Japan, their fellow-OECD member. There are new players, which are not yet and perhaps never will be OECD members, notably in Asia. The Chinese diaspora began the change towards increasing the number of serious world players, though Hong Kong and Taiwan Province of China, and other countries with a

Chinese business class, were joined by the Republic of Korea and Thailand. Then China itself followed suit. From little more than 5 per cent of world exports even in 1980, the Asian tigers and dragons plus China moved up to nearly 15 per cent by 1994; since the mid 1980s the GNP of these countries has been growing at almost three times the rate of the OECD countries.

More importantly, the new Asian economies, or at least their political spokesmen, show no sign of emulating European ways. *The Asia That Can Say No* is the title of a book by the Malaysian prime minister, Dr Mahathir Mohammed, which sets out a 'policy to combat Europe and America'. (The title is actually adapted from the earlier Japanese best-seller by Shintaro Ishihara, 1991, *The Japan That Can Say No*.) Dr Mahathir's thesis is simple; it has often been propounded by Senior Minister Lee Kwan Yew of Singapore and other spokesmen of Singapore's government. It is that Asia can compete with anyone in world markets without abandoning its values. No bricked-up inner cities, no underclass, no drugs and no crime for Asia! Social cohesion — some say Confucianism — will remain the moral basis of life, and will not interfere with economic growth. Indeed, such values may contribute to growth.

How are the dreaded Western values supposed to be kept out? By strong government. Authoritarianism is not totalitarianism. Authoritarian rulers will not brook active opposition, but they leave people alone as long as they do not attack the powers that be. Law-abiding citizens who assiduously attend to their own affairs and otherwise live inoffensive private lives need not fear the wrath of their leaders. The permanent and total mobilization of all by the state which characterizes totalitarian régimes will not happen. Among other things, it would be incompatible with a successful modern economy. But those who criticize government for its unaccountable power, those who use their freedom of speech to expose nepotism, those who dare put up alternative candidates at elections, are in trouble. The limits of civic freedom are tightly drawn.

Is this, then, the alternative with which modern societies are faced: economic growth and political freedom without social cohesion — or economic growth and social cohesion without political freedom? Is there, after all, an alternative to the Western model, equally viable, and more attractive to some, though unacceptable to others? More and more people in the OECD world think so. Many businessmen like the Asian model, and conservative politicians from Margaret Thatcher to Silvio Berlusconi follow suit. Asian values have become the new temptation, and political authoritarianism with them. Abandon the American model, suggests the new wave, and look to Asia for a new model of how economic progress can be combined with social stability and conservative values.

The story is not as new as it sounds. Under different names it has accompanied modern economic development for over a century. After all, was not Imperial Germany a case of the combination of the industrial revolution with an authoritarian régime? And did not Germany grow out of this fallacy?

For countries which embrace modern economic ways before their societies have become civil and their polities democratic, the temptation of authoritarianism is great indeed. (The same is actually true for countries which attempt the economic and political transition from the tyranny of a leader or a party to open society at the same time.) In the early stages, the creation of a market economy invariably requires sacrifices from the people subjected to it. It requires what is called deferred gratification, or in economic terms savings — that is, investment before consumption. People will have to work hard for low wages and tolerate miserable working and living conditions before their countries manage to turn the corner and join the developed world. Such sacrifices are rarely if ever made voluntarily. Even Max Weber's thesis of the usefulness of Calvinism — an early analogue to today's Confucianism — for capitalist progress begs the question of whether it was religious beliefs or authoritarian governments which made people forgo the fruits of their labour. It is hard to think of an example where ascetic values were not reinforced by strong secular powers.

But for the most part, the combination did not last. Capitalism itself changed, to be sure, from saving to spending and on to borrowing. As it progressed, however, society and politics also changed. Increasingly (so the West would like to believe) people demand a share of the wealth they produce; they also want to be masters of their own lives. They want to travel and watch television and choose their own neighbours. They want to have a say in their affairs, a vote, the right to form associations, the possibility to tell a government to go away. Civil society and political liberty follow economic development if they do not precede it. But do they?

The test of history rarely yields unambiguous results for such large theories. Still, this particular one is being tested every day in Asia, and particularly in Japan. There are signs that Japan will go down the Western route, *mutatis mutandis*, which is one reason why Asian leaders criticize it almost as much as the United States. But theories can be wrong; once again there is no inexorable march of History. It is possible that a new Asian — essentially Chinese — balance will be found which combines world competitiveness in economic terms with a social cohesion that is traditional rather than civil, and with authoritarian political régimes. It is also possible that the example will affect European leaders and voters, and that a growing number will wish to go down a similar route at the risk of abandoning some of the cherished rights and liberties of the European and North American tradition. The temptations of such authoritarianism are considerable. They are likely to arise in many policy areas. To mention but a few: integrating the young into society is no longer easy. Where families fail, schools cannot succeed. Labour markets are not exactly waiting for newcomers. Many young people begin to drift and to embrace unsocial behaviour. People want to see them disciplined. Some kind of national service is only a beginning. Someone should be able to tell them what to do, people may say, and to punish them if they do not obey.

Punishment is also the main demand of those who speak incessantly about law and order. The caning of an American youth (who had vandalized cars) in Singapore has led to official protests but much private gloating in the West. The cane, it is said in bars and pubs around Europe, should be brought back in any case. The police should be given greater powers. Life in prisons should be really hard. The death penalty needs to be reintroduced.

The welfare state needs to be reformed, which cannot be done without hardship. But such hardship, people think, should hit first the scroungers who live on other people's money without contributing themselves. If people do not want to work, they must be made to do so, or not get anything. Parents who do not look after their children must, if necessary, be forced to do so. All too often liberty has become licence. The behaviour of people in public is, many think, disgusting. Unkempt men drinking beer in public places, half-undressed girls cavorting about, no one paying respect to the elderly or the infirm — not surprising perhaps in view of the media and the tabloids, but it needs to be stopped. Then there is the invasion of foreigners. In state schools local children have become a minority, and teachers offer prayers in several religions. Somehow this mess has to be sorted out, and England returned to the English, Germany to the Germans, France to the French ...

It would be only too easy to go on. Nor are these the most extreme demands for change. One could probably quote a respectable academic author to give reasons for every one of them. They all add up to the demand for a régime which is less tolerant, one which enforces values at the risk of violating civil rights as we have come to know them, and one which cannot simply be removed by an electoral misfortune. It would be an agreeable side-effect, in the eyes of those whose views are represented here, that such a government could guarantee economic competitiveness without having to listen to disgruntled trade unions, to special-interest groups from Greenpeace to animal rights campaigners or local environmentalists, or indeed to political parties, which are unpopular whatever their persuasion may be.

It is not easy to assess the strength of authoritarian sentiments of this kind in the OECD world. Not everything said by a businessman or a taxi driver in a moment of exasperation amounts to demands for a régime change. There are signs that some of the solutions hinted at will be promoted by perfectly law-abiding democratic forces. Some suspect that law and order will dominate the political agenda to such an extent that perpetrators are bound to face harsher treatment (whatever that means precisely). It would not be surprising to see the death penalty reintroduced in countries which abolished it for good reasons. Certainly, the time of adjustment to global competitiveness with its economic cost for many, social disintegration with the attendant discomforts and pains, and lack of confidence in traditional political parties and leaders, tests the ability of democracies to promote change without violence and without violation of the rule of law.

SOME MODEST PROPOSALS

While we wonder about risks, we must not forget to think about solutions. The point of this argument is now clear. We want prosperity for all, and that means acceptance of the needs of competitiveness in global markets. We want civil societies which hold together and provide the basis of an active and civilized life for all citizens. We want the rule of law and political institutions which allow change as well as critical discourse and the exploration of new horizons. The three desires are not automatically compatible. The challenges of globalization require responses which threaten civil society. The onset of anomie gives rise to the return of authoritarian temptations. These are made more compelling for some by their perception of the Asian model. So what can be done to preserve a civilized balance of wealth creation, social cohesion and political freedom?

One feature of the Popperian approach to problems is that it shuns comprehensive solutions. Whoever claims to have the answer to every question answers none. Total solutions will aggravate rather than improve matters. This means, however, that viable answers are bound to look woefully inadequate in terms of the dimension of the problem. The six suggestions discussed briefly below — which are introduced in order to give some substance to the analysis — all fall into this category. They are a beginning, no more.

First, we have to change the language of public economics. It is actually remarkable how unquestioningly we have all adopted, in public discourse, the concept of 'economism', while the leading economists themselves have moved away from it. For governments, to say nothing of international organizations, GNP growth is still the fetish. Sometimes nowadays they are surprised if growth remains 'jobless' or even 'voteless': even with growth at three per cent, few new jobs are created and the opinion polls continue to show the government in a dim light. Something is missing; some call it the 'feel-good factor'. Yet welfare has long been a concept used by economists. The inventors of GNP accounting, great men like Samuelson or Arrow, speak of the limitations of this yardstick as much as its uses. In a recent book, Partha Dasgupta (1993) contrasts destitution (his main subject) with 'well-being', even 'social well-being'. Like Amartya Sen before him and Meghnad Desai after him, Dasgupta tried to introduce measures for aspects of well-being not included when we speak of gross national product — measures of human rights for example, or of democracy. Clearly, wealth — sustained wealth — is more than per capita GNP. Whether we want to spend much time and energy to produce one complex measure of wealth, or well-being, may be an open question. The answer is probably no. But governments and international organizations should be encouraged to add certain other information every time they produce GNP figures; information about trends in inequality, about measurable opportunities for people, and about human rights and liberties.

A proper wealth audit should take the place of a single simple and often misleading figure.

Second, the nature of work is changing. The straight course of a career for life will be the exception rather than the model. Over a lifetime, people will have been in and out of work, employed full-time or part-time, in training and retraining. In some ways the experience of women will become the general norm, to the chagrin of many young men. Such changes have many implications, for example with regard to the link between social entitlements and employment; entitlements must not be tied to particular jobs. However, this whole transformation can work only if everyone has, at an early stage, gained some experience of the labour market. The new world of work makes it imperative that proper provision be made for young people to pass through a phase of vocational training which is closely related to real jobs and ends up in a period of regular employment. Education does not solve all problems; people understandably ask where it is leading them. But education linked to employment at the critical age — sixteen to nineteen, or so — provides a basis of experience and of motivation which can sustain people through a lifetime of changes. Conversely, if the early opportunity to encounter the uses of education and the constraints of the labour market is missed, much if not all is lost. Thus we should look at best practice in this area and try to apply it more generally.

Third, the truly disadvantaged — the underclass — present an almost unmanageable problem. Clearly, just offering opportunities to those who have fallen through the net is not enough. People will not take such chances without much more inducement. It is unpopular to speak of motivation as an obstacle to the return of some of the truly disadvantaged to the labour market and the wider society; yet it is a fact that many have become indolent and accustomed to a life at the margin. Everything that can be done to include the excluded must be done. Yet the more critical task is another one. In the words of the British parliamentarian and reformer Frank Field, it is to cut the supply routes to tomorrow's underclass. We may not be able to do enough for those already excluded, but we must do enough to prevent another generation from having the same dismal experience. In part, vocational training for all may do the trick. Possibly some form of generalized social service would aid the integration of all into the values of a changing society. Community building through housing and the creation of public spaces might help. The issue evidently needs further exploration. But turning public attention from the remedial to the prospective, from helping today's underclass to preventing the emergence of tomorrow's, would itself help. Advocating international conferences is always a bit of a cop-out; but some agency or other would surely find it attractive to try and bring together expertise and imaginative thought on this subject.

Fourth, globalization means centralization. It individualizes and centralizes at the same time. Intermediate agencies and instances — indeed civil society — are to some extent obstacles of straight globalization. There is a sense in which competitiveness in world markets helps destroy communities. But this does not have to be the case. It is possible to counteract the simultaneous pressures towards individualization and centralization by a new emphasis on local power. The word 'local' is deliberately chosen. Nations within nations — like Wales, or Quebec, or Catalonia — do not have the same effect. They may contribute to a general sense of belonging, but as a principle of social and political organization they divide and produce unhelpful rigidities. (The so-called Europe of Regions is, from this point of view, at best an irrelevance and at worst a mistake.) Local communities, on the other hand, can provide a practical basis for vocational training, for small and medium-sized businesses, for personal involvement and participation, for strengthening the public domain — in short, for civil society — without detracting from economic imperatives. Some countries like Switzerland, and some parts of Germany have much experience in this regard; France and Britain, on the other hand, have suffered from ignoring the potential of local power. As always, there is no general blueprint for sustaining local communities, but some form of political identity, from revenue-raising powers to elected mayors, undoubtedly helps.

Fifth, local power is but one factor in the wider concept of stakeholder economy. Some economists, notably in the United States, do not like the concept; they think that shareholders are the only stakeholders and that the actions of shareholders keep businesses on the straight and narrow. Leaving vast cultural differences — as well as the fact that most shareholders nowadays are institutions — aside, the point about stakeholders is that (contrary to shareholders) they cannot put their interest in companies up for sale. The workforce, the local community, but also banks and even suppliers and buyers are, as it were, stuck with the companies to which they are committed. This can be regarded as an undesirable rigidity only in an inhuman world in which it does not matter whether firms are bought or sold, taken over, merged, extended, reduced, or closed as long as the shareholders get a maximum yield for their investment. In truth it does matter. What is more, competitiveness is not increased by lack of commitment, especially if companies choose to go down the high-skill rather than the low-pay route. Reliability and predictability have their own value in business relations across the globe. Recognition and involvement of stakeholders is the practical answer. This can be brought about in many ways, from works councils for employees to the involvement of banks in investment decisions, from business participation in school boards to the activity of local chambers of commerce. Moreover, organized stakeholder relations are only a part of the story; it is attitudes that matter, an awareness of connections, and a commitment to concerns which, in the end, serve most people best.

Sixth, little has been said in this paper about the role of governments. Acceptance of the fact that in the global market-place the actors are transnational companies, and a preference for the creative chaos of civil society, seem to leave governments out. Yet they clearly are not out of the picture. Nor are they simply the guardians of the rules of the game, as some liberal theorists want it. At the very least, governments set the tone for the economy and for society more generally. Beyond that, governments have a special responsibility for the public domain. Public services require by definition government involvement in funding and administration. Much depends on how such involvement is expressed, both in terms of the value placed on service as a human activity and in terms of the organization of public services. It is quite possible that some OECD countries have gone further in using the public service model than they can afford, or even than is good for the quality of service. It is also possible that in reacting to the experience, some have introduced so-called business values into the public sphere to a point at which both the intended service and the general readiness for commitment suffer. A new balance needs to be found. Health care is quite likely to provide the main example, given its importance for individuals, its cost, and its location on the borderline of global constraints and local or even national opportunities.

This list leaves out many issues which need consideration. Above all, it leaves undecided the critical question of the institutional — one might almost say, the geopolitical — response to the challenges of globalization. Regional blocs of some sort may well be where the world is headed. The cultural debate underlying the project which forms the core of this paper underlines the possible significance of such blocs — Asian values, European values and all that. It is, however, a key part of European — or perhaps OECD — values never to lose sight of the truly international (that is to say the universal) nature of the project for the next decade. We are talking about prosperity for all, civil society everywhere, and political freedom wherever people live. This means that in the end we are not concerned with privileged regions but with one world and its appropriate institutions.

REFERENCES

Albert, Michel (1993) *Capitalism vs Capitalism* (trans. P. Haviland). New York: Four Walls Eight Windows.
Dasgupta, Partha (1993) *An Enquiry into Well-Being and Destitution.* Oxford: Clarendon Press.
Hirschman, Albert (1970) *Exit, Voice, Loyalty: Responses to Decline in Firms, Organizations and States.* Cambridge, MA: Harvard University Press.
Ishihara, Shintaro (1991) *The Japan That Can Say No* (trans. F. Baldwin). New York: Simon and Schuster.
Schlesinger, Jr., Arthur (1992) *The Disuniting of America.* Knoxville, TN: Whittle Direct Books.
Veblen, Thorstein (1915) *Imperial Germany and the Industrial Revolution.* Reprint (1984) Westport, CT: Greenwood Press.
Wilson, William Julius (1987) *The Truly Disadvantaged: The Inner City, the Underclass and Public Policy.* Chicago, IL: University of Chicago Press.

Palace Fundamentalism and Liberal Democracy: Oil, Arms and Irrationality

Fatema Mernissi

Islamic fundamentalism is usually perceived by Western liberal democracies as something not only alien to, but also entirely incompatible with, their philosophical and ethical foundations. Often, though, they fail to make even the elementary distinction between Islamic fundamentalism — an authoritarian ideology and political system, which sacralizes hierarchy and repudiates pluralism — and Islam as a religion and a culture. Thus the incompatibility between Islam and the West has been promoted, since the fall of the communist camp, as the principal field of conflict and lurking danger in the next century.

After making the necessary distinction between Islam as a culture and fundamentalism as a political ideology, I would like to suggest that the liberal democracies in fact have a history of promoting Islamic fundamentalism; and that, in particular, they have made extraordinary profits from Saudi fundamentalism. The internationally overwhelming role of Saudi Arabia as promoter of a kind of aggressive 'petro-fundamentalism' — with its primitive messages of obedience (*Ta'a*), intolerance, misogyny and xenophobia — is inconceivable without the liberal democracies' strategic support of conservative Islam, both as a bulwark against communism and as a tactical resource for controlling Arab oil.

Saudi Wahhabism (named after its preacher Mohammed Ben Abdelwahab, 1703–92) is, by the standards of many Moslems, one of the most fanatical sects of extremist Islam. It insists upon a return to the 'ideal' customs of seventh century Arab desert tribes and considers everything 'added' since the prophet's time to be a foreign perversion — including all scientific and cultural achievements (with their Hellenistic and Persian components). Beginning with the alliance between the preacher Abdelwahab and the warrior Emir Mohammed Ibn Saud in 1740, Wahhabism unsuccessfully tried to invade neighbouring areas. It was halted and crushed by the Ottoman Turks at the beginning of the nineteenth century.

After complete marginalization for more than a century, Wahhabism re-emerged with the discovery of oil, becoming a trump card in the energy and cold war strategies of the liberal democracies. Fanatical Wahhabism proved to be an extraordinary machine for manufacturing 'certainty' in world politics, since it concentrated control over one of the planet's most important sources of petroleum, and the major asset of 230 million Arab citizens, in the hands of one prince and his close court.[1]

Here we see one of the most puzzling marriages of the century: the bond between a fanatical creed and the most modern liberal states. How is this possible? How can liberal democracies oppose democratization in the Arab world? How can liberal democracies support authoritarianism and tyranny? This brings us to the almost unthinkable question: Can liberal democracies be irrational? Concern has been increasing for decades within these countries that the growth and consolidation of a very ambiguous managerial corporate power interferes with the fundamental principles of pluralism and democracy,[2] but this has not seemed to shake the average Westerner's

---

1. This is well described in a recent *Financial Times* article on the Kingdom's budgetary troubles: 'Saudi Arabia's economic policymakers are busy preparing the annual January 1 budget statement, the Kingdom's main public declaration of economic policy ... The "situation" is the worst slap Saudi Arabia has faced since the oil price collapse of the mid-1980s. There is an unprecedented tightness in public coffers — the product of the 55 billion dollar bill for the Gulf War (Desert Storm), subsequent costly military purchases, unhelpfully soft oil prices and the cumulative effects of a decade of high spending ... [H]ow King Fahd weighs such factors in making the budget decision — for the ultimate decision will be his — is unknowable outside his close court ... Only on January 1, therefore, will anyone outside the royal court know quite how serious the Saudi government believes its present economic difficulties really are.' (Mark Nicholson, 'Saudi budget may have to reflect some harsh realities', *Financial Times* 18 November 1994, p. 5.)

2. 'At the heart of the debate is an attempt to evaluate the impact of corporate power on individuals and society. On the one hand, we are afforded apocalyptic visions and dark warnings of tyranny, domination and oppression. On the other hand, we find images of utopia and promises of an organizational society without discontents ... On the face of it, the argument for tyranny would seem to have some merit. In its attention to the formulation and dissemination of ideology, Tech management indeed resembles Big Brother ... [The] facts seem to support the critics' claim that the modern corporation is fast becoming — if it has not already become — a monstrosity' (Bendix, 1956: 339); see also Kunda (1992). Yehuda Shenhav makes a convincing argument on the authoritarian concentration of power which lies behind the concept of 'certainty'; see Shenhav (1994).

strong belief in the pretension of the society to rationality and respect for the individual.

On the contrary, many people in the West would tend to reject as absurd the idea that their government supports Saudi palace fundamentalism, because this destabilizes the comfortable duality according to which the West is rational and progressive, and the East is a dark hole of irrationality and barbarism. Furthermore, many Westerners think it normal to believe that Islam is irrational; and when they say 'Islam', they mean not only the civilization with its religion and cultural heritage and philosophical world-view, but also the entire population, regardless of class, sex, ethnic or economic interests. Their word 'Islam' refers to an indiscriminate magma of people who have the same interests and share the same fanaticism; and their unquestionably popular slogan is 'Islam is Irrational'. But the game becomes more interesting if we include the liberal democracies in the discussion.

IS ISLAM IRRATIONAL? ARE THE LIBERAL DEMOCRACIES RATIONAL?

Liberal democracies, as Francis Fukuyama describes them, leave an Arab woman bathing in dream-like envy. According to Fukuyama, they are uncompromisingly ethical and universalist: 'The chief psychological impera-tive underlying democracy is the desire for universal and equal recognition ... Only liberal democracy can rationally satisfy the human desire for recog-nition, through the granting of elementary rights of citizenship on a universal and equal basis' (Fukuyama, 1993: 6). Many experts like Fukuyama and Samuel Huntington go on to predict that, after the fall of the communist bloc, the next challenge to universalist liberal democracies will come from authoritarian regimes in general and theocracies in particular (see Hunting-ton, 1993). Islam, unlike Confucianism, is singled out as the enemy most totally incompatible with liberal democracies' philosophies and interests.

The problem here is that equating Islam with irrationality immediately turns the world's 1.2 billion Muslims into potential enemies. Since creating such masses of enemies is a bellicosity smacking more of irrationality than of cold analysis, and since many Western intellectuals have produced 'scientific' and 'philosophical' grounds for sustaining this crusade in academic circles, while neo-fascist mobs attack Muslims in many European cities,[3] I suggest that we try to reformulate the question.

---

3. Jurgen Habermas points out that 'Beyond the street attacks of extremists, it is their successes in parliaments which is shocking and constitutes a threat for liberal democracies' ethical foundations ...' (Habermas, 1994: 124). On concerns raised by neo-fascist violence against Muslims in Europe, see the report of the conference on Right Wing Extremism and German Democracy, organized by the Aspen Institute of Berlin, and held on 18–19 June 1994.

If we define irrationality as basically the intolerant behaviour of a person or a system believing in certainty, and therefore holding that there is only one truth and that those who think otherwise ought to be suppressed, then rationality could be defined as the opposite — as what Ralf Dahrendorf has called the 'ethics of uncertainty'. 'The ethics of uncertainty are the ethics of liberty, and the ethics of liberty are the ethics of conflict, of antagonism generated and institutionalized' (Dahrendorf, 1965: 247). Within this framework, we can assume without being completely unrealistic that there must be in the Muslim world some individuals and some institutions (banks, firms, factories) that operate rationally and see their survival as vitally depending on the institutionalization of conflict. Such would certainly be the case for individuals belonging to religious and ethnic minorities (Copts in Egypt, Christians in Sudan, Kurds in Iraq, Berbers in Algeria and Morocco, and so forth), for women who suffer from official legal discrimination, and for free-thinking intellectuals, or simply for individuals who have interests that conflict with the ruling elites.[4]

We can also assume, without being too unrealistic, that some Western citizens and some institutions, although belonging to liberal democracies, might identify a despotic irrationality (as defined above) as suitable for the pursuit of their interests. We hear daily about scandalous aberrations in the ethical rules regulating the liberal market and the political systems of representative democracy. Corrective devices exist in the economic and political systems of the latter precisely because irrationality is assumed to be a possible choice for some individuals.

Assuming that citizens of liberal democracies are rational at all times and in all situations eliminates the dimension of 'uncertainty' so essential to

---

4. The situations of minorities and women are probably the best indicators of secularization in Muslim states. They ought to be adoped as pertinent measures for gauging the balance between rationality and irrationality in these countries, as well as for judging their achievements regarding the establishment of human rights and a civil society. Minorities, women and slaves were the three groups that constituted a challenge and a limitation to Islam's claim to universality and equality. These three groups were the object, in the traditional Muslim state, of special legal dispositions which managed their special inferior status within the Muslim Umma while protecting them as full-fledged human beings.

    This is why most of the debate on 'democracy' circles endlessly around the explosive issue of women's liberation and also why a piece of cloth like the veil is loaded with symbolism and ideological conflict provoking constant violent clashes within and now outside Muslim territories. The heated French debate on *le foulard Islamique*, the Islamic scarf adopted by adolescent Muslim schoolgirls in Paris suburbs, shakes the Republic of France daily and splits its intellectuals and politicians, because it not only confronts France as a liberal democracy with its 'unconscious', irrational, racist inconsistencies, but also forces the Muslim population to discover and deal with its own inconsistencies and clashing aspirations. In this sense, the *foulard Islamique* is far from a trivial secondary issue: it is a highly pertinent symbol upon which both supposedly universal European liberal democracies and their Muslim minorities are testing their claims to rationality, and discovering the limits of those claims.

rationality and smacks of a Muslim Imam's fiat. But once we assume rationality and irrationality to be possible in both liberal democracies and Muslim countries, we can reformulate the question as follows: How can we increase the scope of rationality in the Muslim world? Of course, another question follows immediately: Who are 'we'? 'We' could be any individuals, regardless of culture and nationality, who are interested in nurturing the chances of rationality in the next century; and this 'we' represents millions of lovers of peace and justice within and outside the Muslim world. In fact, the most difficult of all intellectual enterprises today is the attempt to separate 'East' from 'West'.

Focusing our attention on how to increase rationality in the Muslim world has at least a few advantages. The first is that it frees us of the racist bias inherent in opening a debate about which cultural group is rational and which is not. The second is that it empowers us — both 'Westerners' and 'Muslims' — by helping us to identify key factors that can increase the chances of rational problem-solving methods and reduce violent outcomes to conflict. By focusing on the people, on the citizens' desire to exercise their free choice, we reveal the political nature of the conflict. If, to justify their budgets, some generals and arms lobbies find it appropriate to blow cultural differences into a Medieval crusade, we should not jump blindly onto their bandwagon, because we might have different interests — such as promoting dialogue, tolerance and global responsibility, which is, at any rate, the ultimate goal of this essay.

Now let us further narrow the scope of the question to the following: How can we increase the chances for rationality in the Arab world, considering the two determinant strategic factors, which are *oil* and *arms sales*? For when one looks at the issue through the perspective of Arab oil, the landscape shifts dramatically, and strange sights begin to appear. The incompatibility between Beauty (ethical liberal democracies) and the Beast (authoritarian Islam) disappears entirely. As in the Tales of a Thousand and One Nights, Beauty and the Beast can then be seen entwined together, Liberal Democracies and Wahhabism in intimate embrace.

What does this mean for the livelihood of people in the Arab world, and in particular for women and minorities who are singled out as sacrificial victims by fanatics? It means repression (not only of women and minorities, but of all political dissent), support for fanatical political movements, and an irrational use of resources which could provide a decent standard of living for all people, both in the Arab world and outside it. Growing poverty and joblessness in the Arab region (and outside it) oblige us to question the way profits from oil are profligately invested in arms. They force us to ask the only question worth asking: How can we change the situation? How can we create a model of oil management that enhances dignity and well-being in both West and East?

THE HIJACKING OF ARAB JOBS BY THE WESTERN ARMS INDUSTRY

Today, the number of unemployed in the Middle East and North Africa is estimated at ten million (Page, 1995). But a look at the national budgets of the Arab states in general, and of Saudia Arabia in particular, shows that Arab money goes to buy arms which create jobs in Los Angeles and France, rather than in Cairo and Casablanca.[5]

The Middle East is a bonanza market for arms sales — the largest in the world. While the arms purchases of developed countries represented one quarter of the world market in 1985, for example, those of the Middle East amounted to 35 per cent — down from 43 per cent in the two previous years. The end of the Cold War ought to have reduced arms supplies to the region. One might also have expected that the defeat of Iraq and the concentration on the peace process between Israel and its neighbours would have led to a substantial reduction in arms purchases, but this has not been so.

An overview of arms sales published in the latest report of the United States Arms Control and Disarmament Agency reveals that while these sales throughout the world are substantially decreasing, the Middle East scores higher than ever, with military expenditures representing 54 per cent of public outlays and 20 per cent of the GNP of these countries (Stork, 1995: 337). The only significant change since the end of the Cold War and the Gulf War is that the United States now tops the list as the most important arms supplier for the Third World, replacing the Soviet Union. America's most greedy client is Saudi Arabia, with yearly arms purchases of 3.5 to 4 billion dollars (*Middle East Report*, 1994: 1).

At a time of tragic global elimination of jobs, associated with a process of seemingly irresistible structural technological transformation, American and French presidents (as well as others from important arms-producing nations) scour the planet to secure employment for their citizens. The transformation of the heads of state of some major liberal democracies into salesmen seems neither a transient feature of current affairs nor an ethically doubtful event. On the contrary, it looks like an important shift in the role of the state in the post-modern global market. A recent issue of *Newsweek*, with a title story dedicated to 'The New Diplomacy: Uncle Sam as Salesman', announces: 'Now that the Cold War is over, Washington is reaffirming to the world that in foreign policy, the business of America is business'. The American Secretary of State, Warren Christopher, is quoted as saying that if 'for a long time Secretaries of State thought of economics as "low policy" while they dealt only with high science like arms control, I make no apologies for

---

5.   As items in the international press confirm: 'La France va vendre a l'Arabie Saoudite deux frégates pour 19 millards de Francs ... La conclusion de ce marché ... représente 45 millions d'heures de travail pour les entrepresises Francaises' (*Le Monde* 23 November 1994).

putting economics at the top of our foreign policy agenda' (Hirsch and Breslau, 1995). But of course American heads of state and diplomats are not the only ones helping their industries to get contracts; the leaders of other major powers do their best as well.

The arms industry lobbies have proved to be particularly influential in the global market. Not only have they successfully resisted the early post-Cold War desire for smaller arms expenditures and a less aggressive society, but they have also managed to switch the debate from peace to job creation, and have thus enlisted the help of local politicians:

> Opposition to change can come from different directions. Arms producers raise the specter of job losses, so they lobby their governments to buy more weapons, provide higher subsidies and give more support to exports. Local politicians fearing unemployment also argue against the closure of factories and military bases. And within the armed forces, officers and soldiers protest being demobilized. (UNDP, 1994: 9)

These interests find a loyal ally in the Saudi Arabian state. The Arab leader who sits on the world's biggest oil resources, and who claims to defend the Muslim Umma's interest, creates millions of hours of work in the Christian West. If, within a ferociously competitive global market, Saudi Islamic fundamentalism can hijack so many potential jobs from the Arab Mediterranean to Northern Europe and the United States, should we not relinquish the idea that this fundamentalism is an anachronistic, medieval religion and start regarding it as a strategic agency to create employment in the unsettling post-modern economy of the West?

The skilful capture of jobs by liberal democracies, through the sale of arms which make the princes of undemocratic Arab states feel secure (at public expense), is one of the causes of the high rates of youth unemployment in the Middle East. It leaves states like Saudi Arabia open to the street fundamentalists' accusation that governments betray the Muslim community; and it creates a virulent internal opposition, which has gained unprecedented visibility since the Gulf War.[6]

In sum, palace fundamentalism like that of Saudi Arabia is a satanically efficient institution for supporting the global oil and arms lobbies. The young executives working for Boeing and McDonnell-Douglas seem more like the 'cousins' and 'brothers' of the Emirs than do young, unemployed Mustapha and Ali, strolling the streets of Cairo in humiliating uselessness. How has this come about?

---

6. One of the reasons for the emergence of street fundamentalism (a popular protest which expresses itself in religious form) is the fact that palace fundamentalism raised hopes that oil wealth would be shared equally between rich and poor. Massive popular support for Iraq during the Gulf War can be decoded as an expression of people's anger toward Gulf monarchies who, claiming to be saviours of the Muslim Umma, promised solidarity between rich and poor but did not deliver. Furthermore the mistreatment of workers in the Gulf, and of Arab workers in particular, is no secret.

THE MARRIAGE OF PALACE FUNDAMENTALISM WITH LIBERAL DEMOCRACIES, AND THE ATTACK ON ARAB SECULARISM

Wahhabism is one of a myriad fanatic reformist sects. It began in the eighteenth century in one of the most culturally marginal parts of the Arab world, the desert of Arabia. After Mohammed Ali, the sophisticated Ottoman ruler of Egypt, crushed Wahhabism, it was forgotten throughout the nineteenth century. When, in the early twentieth century, the feverishly nationalist Arab World resolutely determined to modernize, secularize and renew itself, no one would have guessed that the Saudis had any role to play on the international scene. Were it not for the discovery of oil and the systematic investment by the West in the region's Emirs from the 1930s onward, few today would know where on the map to find that kingdom.

The secularization of Arab culture and the state, promoted by a widely popular nationalist ideology, was well advanced in the 1930s and 1940s, as has been superbly documented in recent works such as Aziz Al Azmeh's (1992) *Secularism in Modern Arab Life and Thought (Al Ilmaniya)*. Al Azmeh has gathered an impressive wealth of data showing that secularism was neither a superficial nor an elitist phenomenon, but represented a profound transformation running deep at all levels of both popular mentality and state culture.[7] It is important to note that religious authorities were integrated into this intellectual renaissance; and many, like Ali Abderazik (1925), were leaders of reformist movements which produced

---

7. I would like to illustrate how deeply secularism ran in Arab society, and the extent to which the enlightened nationalist culture of the 1940s irreversibly changed people's lives. Arabs of my generation were taught that the reason for our humiliating military defeats, starting with Napoleon's conquest of Egypt in 1798, was the advanced state of Western science. My primary school teacher Faquih Moulay Brahim Kettani, a religious authority and a fervent nationalist, who opened a girl's school to teach Muslim women mathematics and foreign languages, told us that when Napoleon's armada of 400 ships appeared along the coast of Alexandria, 'the most dangerous attackers were not the 36,000 soldiers, but a small group of 151 scientists, engineers and scholars, hardly noticed by anyone in the midst of the chaotic invasion of Egypt, then the jewel of the Ottoman Empire. The secret of the West's compelling superiority was that tiny group'. Thus Faquih used to conclude his Friday prayer in our school mosque, which was built in the 1940s, by reminding us that 'training scientific-minded Arab men and women is the best way, if not the only way, to pray to Allah and strengthen his Umma'.

Unveiling women, liberating them from seclusion and educating them, teaching them mathematics and foreign languages, was the nationalist religious authorities' way of engaging in a Jihad, or holy war, against ignorance. To serve Islam and his God, Moulay Brahim created one of the first mixed primary schools in the midst of the ninth-century medina of Fez. He imposed on us children — girls and boys alike — a tough discipline: Western sciences and the French language in the mornings, Quran and Arab history and poetry in the afternoon, with prayers in between. If not for petro-fundamentalism, how can we explain that now, fifty years later, educated and unveiled women and male intellectuals are the targets of terrorist violence purporting to be religious?

interpretations of the Quran and Hidith accommodating democratization and secularization.[8]

A pertinent indicator of this advanced secularization, accompanied by the appearance of a dynamic civil society, is Egypt's strong feminist movement, animated by Hoda Sha'raoui, an aristocratic beauty who, between 1923 and her death in 1947, managed to influence the ruling elite, get the attention of cultural circles and lead street demonstrations, and who, within her own lifetime, finally obtained changes in marriage law. She also managed to cast off the veil, write, have children and enjoy the unbending support of a loving husband.[9] In Egypt, the famous fundamentalist brotherhood (Jama'at Ikwhan al-Muslimin) was born in the same decade (1928–36) as was Arab feminism. But they co-existed peacefully, with both feminists and fundamentalists stating their visions of the future and defending their ideas side by side. Pluralism was in place in Egypt in the 1930s. The Egyptian feminist movement deserves part of the credit for the fact that the Arab League charter of the 1940s granted women the right to vote, the right to education and the right to work.

Unfortunately for the secular forces of the Arab world, commercially significant quantities of Saudi oil were discovered in 1938. In the following years, oil production increased at an incredible rate, from 30,000 barrels a day in 1943 to 1.2 million in 1960, 8.2 million in 1974 and 10.5 million in 1980 (Corm, 1988: 97). From the beginning of this period, Saudi palace Islam was courted as a way of ensuring 'certainty' in the capitalist market. But it was the Cold War that transformed the romance between the Saudi princes and the oil companies into a stable marriage between Western Liberalism and Saudi fundamentalism.

Saudi Arabia was assigned a new role in the 1950s and 1960s: to fight communism. Arab nationalism, incarnated by Nasser — proudly modernist and devoted to the idea of an *inbi'ath*, a genuine renaissance of the region as an independent actor — was perceived by the West as a dangerous potential ally of the communists.[10] Nasser became the target of Western hostility, expressed in the joint British and French attack on Egypt in 1956. Saudi

---

8. Ali Abderazik's book, *Al-Islam wa Ucul al-Hukm* (*Islam and the Foundations of Power*), published in 1925, is still attacked with fury by fundamentalists. Abderazik stresses the secular nature of the Moslem leader: 'The Caliphate or the Big Imamat is not an institution based on religious creed nor is it a system justified by reason, and all the supposed evidence put forward to justify one or the other does not stand up to examination' (quotation from 1988 edn., p. 148).

9. See Badran's (1986) biography of Hoda Sha'raoui. The best source on Arab feminists is still a contribution by a man, Omar Kahhala, an Egyptian intellectual who compiled a three-volume 'Who's Who' of famous women, listing hundreds of early Arab and Turkish feminists (Kahhala, 1982).

10. The identification of Nasser (who is considered a hero of the Arab nationalist struggle by my generation) as a satanic enemy of the West is captured in this comment by Professor Peter Rodman: 'In the Middle East, for forty years, we and our friends were in a struggle

money poured into the streets of Cairo, financing publishers of medieval literature as a new driving force of Arab thought. Leftist intellectuals were harassed, and many had to choose between prison and exile. It was then that the Arab world lost a substantial number of its intellectuals, who — ironically — often sought exile in the liberal democracies, where they now live and contribute to academic and intellectual life.

The current post-Cold War upsurge of street fundamentalism must be traced to this monstrous alliance between fanaticism and the oil and security policies of liberal democracies. Professor Richard Dekmegian, American Middle East expert and advisor to Presidents Reagan and Bush, provided a succinct account of this phenomenon in a recent interview published in *Al Ahram Weekly* (24–30 November 1994):

Question: How do you account for the rise of Islamist movements in the Arab region and the Middle East?

Answer: Let us start from the time of Nasser. Through Saudi Arabia, the United States provided direct support to the Muslim Brotherhood, the oldest movement in the area. The idea was to counter the rising tide of Arab nationalism and Arab unity championed by Nasser and by the Ba'ath Party in Syria and Iraq. The West became more and more convinced that Islam was the weapon to fight communism. The United States readily spent six billion dollars assisting the Afghan Mujahedin, who were resisting Soviet occupation. Israel shared similar convictions. The Likud government in the early eighties fostered the nascent Hamas to curb Fatah and the PLO.

Question: Do you believe that the Gulf War fuelled the Islamist movements?

Answer: ... I can say with considerable certainty — since I was an advisor to both the Reagan and Bush administrations — that their policies in the region were misconceived and precipitated the war. It is well known that during his first four years in office, Reagan tried to overthrow the Libyan regime of Gaddafi and to consolidate US influence in Lebanon. Israel had failed to do so, and thus the Americans had their try. The reaction was the emergence of Hizbollah in Lebanon and a series of events: the Iran-Iraq War and the Iran–Contra scandal ... Under Bush and Baker, the United States taxpayer provided millions of dollars to support Saddam, the new Shah of the Gulf. Saddam, however, deceived all parties by invading Kuwait. Bush, who had created Saddam, had to remove him.

---

with Gamal Abdel Nasser and his heirs: military strongmen mouthing socialist slogans and backed by Soviet weapons. This menace and this ideology are now defeated' (Rodman, 1995).

Here, then, we come upon that dark, totalitarian side of 'rationality'. Rationality can be perceived to involve creating 'the capacity for controlling future events and creating a life relatively secure from the disruptions of chance. Economic rationality indicates an ability to make reasonable predictions about returns on investments, whereas legal rationality is supposed to provide protection from unpredictable political influences and from sudden changes in the social order' (Chirot, 1985).

Western support for Islamic fundamentalism did manage to further economic 'certainty' during the Cold War era. Palace fundamentalism lessened the element of chance in the market by drastically reducing the number of actors in the Arab political scene. Democratization, it seems, would have multiplied the actors and atomized decision-making by forcing leaders to submit to accountability checks. Thus democracy in the Arab world would have meant an intolerable increase in unpredictability in the oil and arms markets. The enlargement of the political theatre, to allow Arab citizens to express their free will through twenty-two parliaments of Arab states, would have introduced a high level of uncertainty and posed 'a barrier to rational economic behaviour' (Shenav, 1994: 270).

But as Yehuda Shenav has noted, 'increased rationality increases the certainty of one group, but the uncertainty of another' (ibid: 269). As the liberal democracies' economic and political strategies tilted the balance against civil society in the Arab World, they contributed to making the life of the average Arab citizen in general, and the lives of women and minorities in particular, a terrible field of insecurity. People's incapacity to control events in their lives or to change their situation has been one of the main themes in songs, literature and films, and lies behind the anger of young people with states that have become increasingly incapable of doing their job, which should be to defend the interests of the masses.

INTERVENTIONIST STATES AND INITIATIVE-DEPRIVED STATES

During the Cold War, then, Western liberalism perceived and approached conflict, pluralism and democratization in the Arab world as a threat to its interests. After the Cold War, things seem more ambiguous. The possibility of shifting alliances is more than probable, especially because the viability of the Middle East peace process and the security of Israel depend upon the existence of a democratic and tolerant Arab world. There is an emergent strategy to promote stability which is based on nurturing democratic Arab forces and strengthening civil society; and this gives the old lovers (Liberal Democracy and Saudi Princes) a vitally important opportunity to change their relationship. The emergence of a highly educated and sophisticated younger generation in the Saudi Kingdom, in particular, creates a possibility for beginning the twenty-first century on a new footing. But this possibility will be affected by the nature of the Arab state and the need for this very ambiguous entity to redesign itself in a modern form.

With the collapse of communism, debate on the nature of the state has blossomed. An excellent example was the exchange between Francis Fukuyama and Ghia Nodia on the conflict between supposedly universalistic liberalism and nationalism.[11] The debate centres on the question, 'Whose interests does the state serve? Exclusively those of its own nationals, or those of all humanity?'. Fukuyama argues that the liberal state is universalistic and tends to further the interests of all human beings, while nationalist states discriminate against all who do not belong to the 'nation'. In the debate, there are only two kinds of states, and both have the same capacities for intervention. The only difference is the number of people they intend to serve. Nationalism is inferior, according to Fukuyama, because it blocks the freedom of choice of the individual and therefore cannot win in the competition with liberal democracies:

> If, as I have argued elsewhere, liberalism is about the universal and equal recognition of every citizen's dignity as an autonomous human being, then the introduction of a national principle necessarily introduces distinctions between people. Persons who do not belong to the dominant nationality *ipso facto* have their dignity recognized in an inferior way to those who do belong — a flat contradiction of the principle of universal and equal recognition. (Fukuyama, 1992: 24)

Anything which interferes with the individual's free choice, or curtails his/her sovereign right to decide for him/herself, according to this view, is irrational and reduces the potential for creativity of a people.

I consider this position quixotic, not only because liberalism is far removed from treating all Arab citizens as human beings, but also because the way the question is posed diverts attention from a terrifyingly dangerous polarization in the post-modern global society — a polarization between what we could call 'interventionist' states initiating planetary strategies, on the one hand and, on the other, states executing these strategies, without having played any part in initiating them. Taking the oil and arms nexus as an example, we could say that, during the Cold War, liberal democratic states were able to develop planetary strategies, intervene in the affairs of Arab states, and implement these strategies successfully; while Arab states were unable to develop any alternative other than to execute such strategies, even if they revealed themselves to be disadvantageous to their citizens.

Indeed, one reason for the weakness of the initiative-deprived state is the absence of a democratic base — the paralysis of civil society, which is the only instance capable of producing strategies with a broad democratic groundwork, allowing diverging interest groups to negotiate balanced

---

11. See the special issue of the *Journal of Democracy* devoted to this debate (Vol. 3, no. 4, October 1992), and in particular Nodia (1992) and Fukuyama (1992).

solutions for vital problems such as employment and job creation. The more locally focused and democratically produced the strategies, and the broader the range of actors involved in their elaboration, the wider the support they are likely to mobilize and the stronger the interventionist capacity of the state.

Western liberal democracies in general, and the United States in particular, fit the description of interventionist states with planetary strategies. Far from embodying a perfect liberal state representing universal interests (such a state does not, in fact, exist anywhere on the planet), the American political system is inclined toward self-interested protectionism. As the American political scientist James Kurt has pointed out, the privileged connection between the United States and international organizations (such as the International Monetary Fund, the World Bank and the United Nations) endows that country with 'a beyond-the-nation-state interventionist capability' (Kurt, 1993: 13).

This distinction between interventionist and executant states helps to explain the 'conspiracy syndrome' so common among Arab citizens, who complain that the West interferes in their affairs, and that the CIA and 'the Lobbies' are responsible for everything that happens in their lives. To say that such conspiracy theories are imaginary is to ignore the clear discrepancy between the capabilities of Western and Arab states, and to ignore the structure of the oil and arms markets, which are far from a universal paradise in which all human beings are treated equally.

MAKING OUR DIFFERENCES INTELLIGIBLE

It is understandable that a Western intellectual who is neither an arms dealer nor a banker associated with the oil monopolies could feel that his or her moral probity is being attacked, and could become defensive when an Arab intellectual complains about the tragic influence of the liberal democracies in the region during the Cold War. Of course, the average citizen of liberal democracies is not responsible for the atrocities committed against progressive forces by Arab authoritarian regimes during the struggle against communism. But dismissing these emotionally charged perceptions as 'paranoia' is indeed a poor response on the part of those intellectuals whose role in shaping their own countries' policies is more vital than ever before.

Western intellectuals are the target of complaints on the part of their colleagues from the Third World precisely because the former are perceived to be a source of hope for transforming mentalities in a dangerously consumerist and souk-like global market. Intellectuals everywhere are emerging as what Wolf Lepenies calls 'translators between cultures'.

> We cannot be satisfied any longer with simply trying to *understand* other cultures. Understanding is an attitude which involves distancing — an attitude adopted toward

cultures that are only taken into account in a very indirect way, cultures with which one is willing to establish contact only in order to improve one's general knowledge. All that has changed. We are now obliged to apply ourselves to the task of rendering our cultures *intelligible*, because we are increasingly compelled to understand each other within the much more immediate context of *living together*. For all intellectuals, this is an enormous challenge. (de Bresson and Kajman, 1994: 2)

The dynamics of oil, arms and fundamentalism are not a bad place to start in rendering the relations between the peoples of the West and the Arab world intelligible. We do not live in separate worlds, but in highly interconnected ones. In the new post-Cold War world, let us rethink the entire approach to economic development and democratization in the Middle East, giving 'stability' and 'security' a different and more positive meaning. Western intellectuals and policymakers can make a great contribution in this regard by adopting a sense of responsibility and commitment to democracy commensurate with their great capacity to control the world's resources. Reversing earlier policies of support for autocratic regimes and nurturing the revitalization of civil society in the Arab world would be a daring and constructive way to step into the twenty-first century.

REFERENCES

Al Azmeh, Aziz (1992) *Secularism in Modern Arab Life and Thought* (*Al Ilmaniva*). Beirut: Markaz Dirassat al-Wahda al-Arabiya.

Abderazik, Ali (1925) *Al-Islam wa Ucul al-Hukm* (*Islam and the Foundations of Power*) (1988 edn.: Beirut: Al Muassassa al Arabiya li-dirassat wa-Nachr.)

Badran, Margot (1986) *Harem Years*. London: Virago Press.

Bendix, Reinhard (1956) *Work and Authority in Industry*. New York: Harper and Row.

de Bresson, Henri and Michel Kajman (1994) 'Un entretien avec Wolf Lepenies', *Le Monde* 31 May, p. 2.

Chirot, Daniel (1985) 'The Rise of the West', *American Sociological Review* 50: 181–95.

Corm, George (1988) *Le Proche Orient Eclaté*. Paris: Gallimard.

Dahrendorf, Ralf (1965) 'Uncertainty, Science and Democracy', in *Essays in the Theory of Society*, pp. 232–55. Stanford, CA: Stanford University Press.

Fukuyama, Francis (1992) 'Comments on Nationalism and Democracy', *Journal of Democracy* (special issue) 3(4): 23–8.

Fukuyama, Francis (1993) 'Capitalism and Democracy: The Missing Link', *Dialogue* 100(2): 6.

Habermas, Jurgen (1994) *The Past as Future*. Omaha, NE: University of Nebraska Press.

Hirsh, Michael and Karen Breslau (1995) 'USA, Inc.', *Newsweek* 6 March, p. 10.

Huntington, Samuel (1993) 'The Clash of Civilizations', *Foreign Affairs* 72(3).

Kahhala, Omar (1982) *A'lam an-Nissa*. Beirut: Muassassat ar-Rissala.

Kunda, Gideon (1992) *Engineering Culture: Control and Commitment in a High-Tech Corporation*. Philadelphia, PA: Temple University Press.

Kurt, James (1993) 'Toward the Postmodern World', *Dialogue* 100(2).

Middle East Report (1994) 24(6) November–December.

Nodia, Ghia (1992) 'Nationalism and Democracy', *Journal of Democracy* (special issue) 3(4): 3–22.

Page Jr., John (1995) 'The Middle East: How Far from East Asia', note to a World Bank Advisory Board meeting on the Middle East and North Africa, Beirut (March).

Report of the conference on 'Right Wing Extremism and German Democracy', Aspen Institute, Berlin (18–19 June 1994).

Rodman, Peter (1995) 'Arab Democracy/American Democracy', presentation at the Conference on 'Reshaping the Agenda: US-Arab Relations', hosted by the Foundation on Democratization and Political Change in the Middle East, Washington (October).

Shenav, Yehuda (1994) 'Manufacturing Uncertainty and Certainty in Manufacturing: Managerial Discourse and the Rhetoric of Organizational Theory', *Science in Context* 7(2): 267–307.

Stork, Joe (1995) 'Des arsenaux en quete de clients solvables', *Le Monde Diplomatique* January: 327–51.

UNDP (1994) *Human Development Report*. New York: Oxford University Press.

The Future of the State

E. J. Hobsbawm

Virtually all of us are today living in a particular kind of political entity which determines our lives: the sovereign territorial state or 'nation-state'. Some Western countries have lived under such states since the late eighteenth century, in one or two cases possibly even earlier. The dependent territories of the former empires have lived, or tried to live, under them since decolonization. This kind of state has become the universal framework of social development. The future of the state is therefore directly relevant to the subject of the present volume, all the more so because, as I shall argue in this paper, the sovereign nation-state is today entering a new phase. After a period of virtually unbroken advance from the late eighteenth century to the 1960s, it has entered an era of uncertainty, perhaps of retreat. A phase of state development which lasted for about two centuries is now at an end. What the future will bring is not clear because, as I shall also try to show, no alternative devices for fulfilling some of the social tasks hitherto carried out by states have yet been devised.

SPECIFIC CHARACTERISTICS OF THE TERRITORIAL STATE

The type of political entity in which we live has a number of historically specific characteristics, which differ from those found in past forms of political organization. In the first place, it consists of a (preferably continuous and unbroken) territory, separated from other states by clearly defined lines (frontiers or borders) demarcating the area under the power of one state government from that under the power of another.

Second, this territory is sovereign, which is to say that within it no authority other than that of the local state is recognized, except by the unforced agreement of that state (by a negotiated treaty, for example). 'Extraterritoriality' is thus a synonym for the absence of the local state's power. The imposition of a superior authority against the will of the local state — such as a military conquest — is an act of force which brings sovereignty to an end, at least temporarily. One reason why nation-states have a strong preference for a single continuous territory is that a discontinuous territory — as in the case of the Gaza strip and Jericho — means that some other power has control over the communications of the state with parts of its territory.

Third, within its territory the state has the monopoly both of law and of the powers of coercion, except insofar as these have been willingly renounced, as by the member-states of the European Union which accept for certain purposes the precedence of European law over national law. The state's authority extends to all who are present on its territory; and while on that territory, again with minor exceptions such as diplomatic immunity, all persons are, to varying extents, subject to that state's power and to no other. Nevertheless states generally recognize a distinction between citizens or subjects of the state, properly speaking, and temporarily or permanently resident strangers, generally citizens or subjects of some other state. These have only limited obligations to the local state while resident on its territory and, usually, more limited claims upon it, or rights within it. On the other hand, the obligations of a state's citizens towards it are, or have tended to become, unlimited, up to and including the obligation to kill and let themselves be killed on its behalf. The state has the right to prevent any strangers from entering its territory, or its citizens from leaving it. States may choose to exercise these powers to a greater or lesser extent, but all of them possess these powers.

Fourth, the national state rules its citizens or subjects directly and not through intermediate authorities, subaltern but to some extent autonomous, as was the case in feudal societies or certain kinds of traditional empires. There is a fundamental difference between such 'indirect rule' (which did not necessarily exclude the occasional direct intervention of the supreme authority at all levels) and the modern functional differentiation between central, regional and local government.

In the typical territorial state local government supplements but does not replace direct rule by national government, even when the former is not actually dependent on the decisions of national government. State law is the only law at all levels, or at all events is superior to local or regional law where it comes into conflict with these. In centralized democratic states, local or regional elections have no constitutional bearing on the determination of the national authority, which is directly elected by the body of citizens. At least some matters concerning the lives of the country's inhabitants are directly administered by agents of the national government. For most of the period between the late eighteenth century and the 1960s, there was even a tendency for most federal or confederal states to shift the locus of effective power from state to federal or central government, a tendency which is still continuing in the greater part of the Western hemisphere. For the purposes of this paper it is not necessary to consider the exceptions to this generalization.

Fifth, direct government and administration of the inhabitants by the central authorities of a nation-state implies a certain degree of standardization or even homogenization in the treatment of the inhabitants. If, say, monogamy is established by state law, then polygamy cannot be legally practised in the state (as the Mormons in Utah discovered when joining the United States). One of the first monuments to the concept of the territorial state was the sixteenth century German principle *cuius regio, eius religion* (religion goes with state territory). This principle was later abandoned, though twentieth century experience demonstrates that states may decide once again to impose universal orthodoxies on their citizens and subjects. However, while in theory there is no limit to the imposition of such uniformity, in practice the extent to which states insist on it varies enormously with time, place, subject and the state's concerns.

Finally, while a nation-state composed of people without political rights or positive participation in its affairs is possible, the heritage of the Age of Revolution has been to turn most states into citizen states, at least in theory. In such cases the state is considered to represent 'the people', and 'the people' to be the source of sovereignty, or at least to give the state legitimacy, most commonly by some form of election or plebiscite, or some other form of public ritual symbolizing the unity of people and state. While this does not mean that citizen-states are necessarily democratic in any sense of the word, it implies that 'the people' of the state is understood to consist of all or most indigenous inhabitants permanently resident on its territory, or at least of the heads of households deemed to represent their families. Increasingly, with political democratization, this has been taken to mean all adult men and women. States which clearly exclude a majority or even a substantial body of their inhabitants from citizen status (as in South Africa under apartheid and in other white settler states) cannot be considered citizen states. Conversely, as in the United States, the mass immigration of foreigners hoping to settle permanently in the country has been regarded as tolerable only on the assumption that most of them would wish to become citizens at some point.

Citizen status implies, at least for certain purposes, such as their treatment by the law, the equality of all citizens. (This is to some extent implicit in the standardization of the country's inhabitants by the state, which has been noted above.) In states built on the principles of the Age of Revolution, equality tended to be seen essentially as a relationship between unattached individual persons who, as citizens, were qualitatively indistinguishable.

Their membership in social groups, communities and other collectivities was regarded as irrelevant, or indeed incompatible with political and legal equality. Earlier forms of political rule had been both inegalitarian and primarily based on such group membership.

Between the era of the American and French Revolutions and the mid-twentieth century, states with the characteristics noted above changed in several significant respects. They extended their range, power and functions almost continuously. Even when states themselves were prepared to leave most of their citizens' affairs to the operations of civil society and economic *laissez faire*, as in mid-nineteenth century Britain, they had at their disposal far greater capacities than the pre-revolutionary absolute monarchies, not least the ability to organize national censuses, postal services and police forces centrally controlled and reaching into every corner of the national territory. In most of the modernizing states, the inhabitants and private centres of power were largely disarmed, and the powers of coercion became in effect a state monopoly. From the late nineteenth century onward, their functions extended into the field of universal state-organized education and the beginnings of state-financed social security. This was to be enormously expanded subsequently. The total wars of the first half of the twentieth century brought states into the planning, control and management of their national economies and the mobilization of all their inhabitants for national service. While there were obvious differences between states dedicated to the maintenance of liberal capitalism and other variants or systems of the economy, the tendency was general. Whether governments were liberal, authoritarian, social-democratic, fascist or communist, at the peak of this trend the parameters of citizens' lives were almost wholly determined by what their states did or omitted to do, except during international conflict, when they were also determined by the activities of some other state(s).

With the democratization of politics in the late nineteenth century, a new and dangerous element was incorporated into the concept of the nation-state. Citizens had hitherto been regarded primarily as a body of adult human beings united by accepting a common law and constitution, and represented by the same (elected) law-giving authority. Since it was obvious that the inhabitants of all but the tiniest states were a heterogeneous ensemble, their ethnic origins, religious beliefs, spoken language or other personal characteristics had nothing to do with their nationality. This was simply defined by their citizenship. The new element was the assumption that the citizenry of a nation-state formed a 'community' whose members were united by a supposed common origin ('ethnicity') and history, by common language and culture, by symbols, mores and beliefs — in short that they were or ought to form a homogenous population. This may be (and was) put in another way: the world consisted of 'peoples' or 'nations', each with its own way of life based on language and ethnicity, and not to be compared with any other; and each was entitled to form a separate territorial state. Systematic attempts to form such homogeneous ethnic-linguistic states have been made from time to time since the First World War. They implied (and imply) the breakup of all large pluri-ethnic and pluri-lingual states and, since humanity is not in fact divided into neatly separable pieces of homogeneous territory, the forcible homogenization of ethnic-linguistic nation-states. The methods for achieving this have, since 1915, ranged from mass population transfers to genocide.

TENDENCIES AFFECTING THE MAJOR STATES AND THEIR FUTURE

The world is at present composed of just under 200 states, of which about 25, with populations of over 50 million each, include about three quarters of the global population. At the other extreme, 71 of the political units treated as 'economies' by the World Bank in 1992 contained populations of less than 2.5 million; 18 of these contained populations of less than 100,000. The present essay is obviously concerned primarily with the major states and their future.

For a number of states the past decades have been an era of catastrophe and disintegration, as in the former communist region and in large parts of Sub-Saharan Africa. In some areas state power and organization have, for practical purposes, ceased to exist. Special reasons can, however, be found for these regional catastrophes. Here we will be concerned with general tendencies which affect even the oldest, most firmly established and traditionally strongest and most stable states. For even many of these are no longer able to carry the load which they shouldered so successfully in the past. Thus even the provision of public and private security for persons and property, on which developed Western states used to pride themselves, can no longer be guaranteed; and consequently the fear of 'crime on the streets' and the demand for 'law and order' have become powerful political forces in a number of Western countries.[1] The rise of transnational entities operating within states, beyond their effective power, also means that the state no longer has sovereign control over what happens on its territory. In fact, states may have to negotiate with transnationals as with other sovereign powers.[2] Moreover, the retreat of the state is indicated by the extent to which functions formerly carried out habitually by public authorities, or even regarded as quintessentially the state's duties, such as the provision of postal services, are today carried out partly by private businesses.[3]

---

1. As an illustration from peaceful and law-abiding Scandinavia: a hotel near Oslo in the early 1990s advertised its facilities for holding congresses and seminars by stressing, among many other things, that its windows were bullet-proof.
2. For example, state power in Spain, or Catalonia, was helpless when faced with the unilateral decision of Volkswagen to stop manufacturing SEAT cars in Barcelona, with serious consequences for the regional economy.
3. In extreme cases, as in countries governed by neo-liberal economic theology, even prisons have been 'privatized'.

Conversely, it has become increasingly evident that the late twentieth century generates problems requiring global action. Their solution lies beyond the powers of single states or groups of states, however large and effective. Global ecological and environmental problems are a familiar example. Thus the state's powers and functions have been undermined by the rise of supranational and infranational forces, as well as by what can best be described as the withdrawal of its inhabitants from citizenship. This does not mean that its present situation can be described as one of crisis, although this has been argued. Supranational forces have weakened the state in three ways.

First, the creation of a supranational (or rather transnational) economy, whose transactions are largely uncontrolled or even uncontrollable by states, restricts the capacity of states to direct national economies. A major reason for the crisis of the social-democratic and Keynesian policies which dominated Western capitalism in the third quarter of the century is precisely that the power of states to set levels of employment, wages and welfare expenditures on their territory has been undermined by exposure to international competition from economies producing more cheaply or more efficiently. The fashionable neo-liberal free market policies of the 1980s have made states even more vulnerable.

Second, the state has been weakened by the rise of regional or global institutions — such as the European Union and the international banking institutions set up in 1945 — to which individual states defer, either because they are too small to engage in effective international competition except as part of a larger bloc, or because their economies (or, more precisely, their public finances) are so weak as to make them dependent on loans given under politically restrictive conditions.

Third, territorial borders have been made largely irrelevant by the technological revolution in transport and communications. A world in which people, with rare exceptions, lived and worked either in one state or in another, has been replaced by one in which they may live and work simultaneously in, or commute between, more than one state, as well as being in constant immediate contact with any part of the globe. It is quite normal today for a person of relatively modest economic situation to be simultaneously a householder and income-earner in two or more states. The sharp distinction between permanent and seasonal, or temporary, emigration, so typical of the wave of intercontinental migration before the First World War, no longer applies in the present wave of international migration. This affects the relations between permanent immigrants and the states in which they have chosen to live, as well as between immigrants and their countries of origin. It also raises the question of social and political rights for people domiciled in a state who are not, and may not wish to be, full citizens of that state. A glance at current politics in Europe, North America, and the successor states to the USSR shows that these are explosive issues.

Since the 1960s a tendency for state centralization to decline, even in hitherto very Jacobin states, has been notable in many Western countries, and the disintegration or transformation of communist and former communist régimes can be seen in the same light. (A similar tendency to decentralize the structure of the great private transnational business corporations could also be noted.) To what extent this occurred for reasons of operational efficiency, to what extent it was a response to regionalist or nationalist demands, and indeed, to what extent such decentralization necessarily weakened the power of central government, remain open questions. Nevertheless, as the rise of active secessionist movements within some of the oldest 'nation-states' demonstrates, its potential for undermining traditional state power is undeniable. This is, even more obviously, true of the fragmentation of formerly centralized communist states. Nevertheless we face the important question of whether the smaller political units so produced, whatever their degree of autonomy or sovereignty, are substantially different from the larger states in anything except size — that is to say, whether or not they are undermined by the same forces that weaken any form of the old state.

There can be little doubt that the links between citizens and public affairs are in the process of attenuation, at least in states with democratic politics, for various reasons. The decline in ideological mass parties, politically mobilizing electoral 'machines' or other organizations for mass civic activity (such as labour unions) is one of them; another is the spread of the values of consumer individualism, in an age when the satisfactions of rising material consumption are both widely available and constantly advertised.

This withdrawal of citizens affects the legitimacy of national governments and their functions, as well as the demands which they can make upon the country's inhabitants. For instance, it may undermine the assumption, fundamental to the operations of twentieth century states, that citizens will accept the raising and distribution of public revenues for activities in the common interest, and according to a criterion of social equity — in other words, that even those who pay more in taxes than they receive in public services and benefits, accept the legitimacy of some degree of progressive taxation and interpersonal or inter-regional redistribution. The state is weakened when it is not identified with a common good, or when only individual advantage and not common interest are recognized — for example, when *any* taxation is seen essentially as a diminution of the individual's purchasing capacity. The currently widespread dislike of 'bureaucracy', 'government', 'excessive state interference' and the like may contain both a justified critique of the uses and modes of state power and the unjustified assumption that any limitation by superior authority of the individual's right to do what she or he wishes, is unacceptable. We need to distinguish clearly between them.

The current situation of the traditional state has led some observers to underestimate the state's continuing functions and powers. This has been encouraged by the recent, but now declining, popularity of ultra-*laissez faire* theories and policies. We must distinguish between what states cannot do, on the one hand, and what they could do if they wanted to. Much of what governments refrain from doing is rejected not because it is ineffective — for instance, economic protectionism and a degree of self-sufficiency can work — but because, for various reasons, governments do not desire it. Governments like that in Britain since 1979 have had no difficulty in running counter to the general trend of decentralization, producing the most centralized administration and the most severe weakening of local democracy in British history. The difference between Bosnia and Northern Ireland — between a catastrophic civil war and a low-intensity conflict with approximately 100 deaths each year over a quarter of a century — is that in one case the state (Yugoslavia) abdicated and in the other it did not. States are weaker than before, but (unless they collapse and disintegrate) they retain very substantial powers. How much and in what way they should use them, are separate questions.

TWO UNPROMISING ALTERNATIVES: *LAISSEZ FAIRE* AND 'SMALL IS BEAUTIFUL'

Nevertheless, all qualifications made, the nation-state's situation is uncertain. There is a search for alternatives. Two of these are unpromising: anarchism, in the currently fashionable form of economic free-market ultra-liberalism; and 'small is beautiful', or the substitution of larger political units by smaller ones.

Although anti-state liberalism has become fashionable on various grounds, it must be rejected. 'Civil society' and the state are not mutually incompatible opposites, but symbiotic. Nineteenth century *laissez faire* liberalism did not oppose the state in itself, without whose guarantee of law, security and the performance of contracts no economically rational free market was possible, but rather some state activities. Adam Smith accepted and recommended state intervention on a considerable scale, since he took it for granted that important public and private functions could not be adequately supplied by the market. The attempt to implement pure free-market policies since 1980 has confirmed their limitations, as the experience of societies of the Soviet type demonstrated the limitations of centralized planned economies relying exclusively (in theory) on state power. Whatever the most desirable balance may be between public and private, state and civil society, government and market, nobody seriously doubts that they must be combined.

It may also be argued, though perhaps not quite so confidently, that the basic problems of the traditional state are not removed simply by reducing its size. Smaller political units are no doubt potentially better than larger ones since the distance between the centre of authority and the citizen is less, and so, therefore, is his or her alienation from those who govern or administer the territory. And yet, is the inhabitant of New York state significantly closer to the governor in Albany than to the president in Washington? The social or political units in which human beings genuinely feel close to authority and able to influence it as individual agents are almost certainly far smaller than all but the tiniest member-states of the United Nations, or the administrative subdivisions of even medium sized countries. Sociologically speaking, they are something like face-to-face communities: workshops or plants rather than firms or industries, smaller cities rather than metropolitan cities, provinces or regions. It is no accident that French politicians, seeking grassroots democratic credibility, still prefer to be identified with provincial towns — allowing themselves to be elected as mayors of usually modest municipalities — rather than with departments or regions. That is why, to use Benedict Anderson's convenient phrase, even the smallest nation is an 'imagined community'.[4] Even that community may separate the citizen from the decisionmakers by bureaucratic layers of varying degrees of impenetrability, especially if it constitutes an independent state.

However desirable it is to diminish the distance between decisionmaking authority and citizens and to strengthen the 'natural units' of democratic politics, the mere decentralization, regionalization or breakup of existing states in itself provides no solution for the current problems of the state, as sketched above; not even for those of democracy. A world composed of nothing but states of the size of Luxembourg or Antigua, or even of city-states like Singapore and Hong Kong, would not be a world in which the lives of its inhabitants would be more determined by the democratic decisions of citizens than is the case in France or the United States, but one in which the decisions affecting their lives would be almost wholly taken by non-state entities beyond the control of voters. Insofar as the problems of states depend on size, they reflect the fact that states are too small to cope effectively with the global, transnational, and certainly supranational scale of the world.

Moreover, both *laissez faire* and 'small is beautiful' tend to make one of the major problems of the world more acute, namely the politically and socially dangerous growth of inequality between regions and classes. That free market policies, uncorrected by public redistribution, create social inequality needs no proof after the dramatic increase in the inequality of incomes in the 1980s. It is less commonly noted that political fragmentation, national or municipal, generally appeals most strongly to prosperous regions which can thus avoid subsidizing poorer ones: to Slovenia and Croatia, as against Bosnia and Montenegro, to Lombardy as against Calabria, to

---

4. 'It is imagined because the members of even the smallest nation will never know most of their fellow-members, meet them, or even hear of them, yet in the minds of each lives the image of their communion' (B. Anderson, *Imagined Communities* (revised edition), London and New York: Verso, 1991).

Catalonia as against Extremadura, to Santa Monica as against south-west
Los Angeles.

THE INDISPENSABLE REDISTRIBUTIVE FUNCTION OF THE STATE

In this respect the state, preferably the large state or a supranational
combination, remains indispensable. It is to this day the main mechanism
for social transfers, that is to say for collecting an appropriate fraction of the
economy's total income, usually in the form of public revenue, and redist-
ributing it among the population according to some criterion of public
interest, common welfare and social needs. How much of their national
income states should collect and redistribute, how these transfers should
operate and by what criteria of social equity, are open questions, but the
need to provide for education, health care, income maintenance, and so
forth is not; nor is the indispensability of some sort of public authority for
these general purposes. Indeed, after fifteen years of systematic ideological
hostility to redistribution, the scale of such state transfers has not
diminished even in states ruled by governments devoted to *laissez faire*
and the shrinking of state activity, such as Britain and the United States. A
number of the traditional functions of government can be carried out by
private agencies, or through the market, but not this. The redistributive
function may be divided between public authorities at different levels, from
local government to the European Union, and perhaps some day a world
authority. It may also be carried out in a variety of organizational ways.[5]
However, at present it is difficult to see its being carried out over any large
territory or population except by or through public authority, or at least
through an agency which, however cost-effectively conducted, regards its
social purpose as paramount.

The redistributive function, already crucial, is likely to become even more
important because trends in economic development will make it so. The
wealth of developed economies continues to multiply, but this growing
wealth seems likely to be generated by a smaller proportion of their
populations. It is often argued, against welfare states, that as the population
ages in an economy, a diminishing percentage of working age has to carry the
growing weight of paying for a rising percentage of non-earners. The same
argument can be made about the growing proportion of young people who,
being in full-time education, are also non-earners. This may suggest that the
economy is unable to afford a continuously rising standard of living and

---

5. Thus social security transfers, which in the West were mainly made through state agencies,
   in the former Soviet Union were, in practice, largely ensured through the citizen's place of
   employment, she or he being theoretically guaranteed employment. Hence the privatiza-
   tion of the economy would leave the country without an effective social security system,
   until one could be constructed.

welfare for all, as it could do in the past. But this cannot be an argument against redistribution. On the contrary, it makes the case for public redistribution. If the proportion of people earning an income directly through the economy (typically as earners of wages/salaries) will fall, then some mechanism for guaranteeing non-earners a share of the economy's income is indispensable. Unless some other non-market mechanism of redistribution is devised, provision for such people will have to be based on, or built around, public authority. For the time being the nation-state remains the most effective authority for this purpose. It may not be ideal for the purpose in a much more globalized world. Nevertheless if compared to global institutions which (for the time being) have no power to diminish international economic inequalities, nation-states (supplemented in Europe by the European Union) are in a position to diminish regional or group inequalities to some extent.

THE UNCERTAIN FUTURE

These are not arguments for the traditional nation-state as such, but for the (preferably large, pluri-ethnic and pluri-regional) nation-state as something that has not yet been replaced adequately by supranational and/or infra-national entities. It is still the best unit we have for the time being, not least from the point of view of democratic politics, for which supranational, transnational and global authorities provide little or no real space.[6] The governments of Iran or China are more responsive to political pressure from below than is a large transnational corporation or the IMF. It would follow that the increasing globalization of world affairs has implications, possibly negative, for the future of political democracy. Nevertheless it is increasingly clear that nation-states will have to be supplemented or in important respects replaced by bodies capable of dealing with the problems of the global environment, the global economy, global demographic movements, global inequalities and, increasingly, the globalization of communication and culture.

How this is to be done is obscure, especially as the model pioneered by the European Economic Community in the 1950s (the future of which is now uncertain) has not so far been followed elsewhere. However, for the reasons suggested above, the territorial state will continue to play a major role in relation to social development. At the very least, a full and planned utiliza-tion of its resources will help to defend its people against the uncontrolled

---

6. Even the European Union, which is gradually removing what has been called its 'democratic deficit', has so far failed to create the basis for an all-European political life. Elections to the European Parliament in all member countries are conducted entirely in terms of national politics, insofar as they arouse any serious interest at all.

impact of global forces on their way of life, and against the catastrophes into which a sudden and uncompensated collapse of the state, as in the ex-Soviet Union, can plunge a large part of the world.

The National Question in Africa: Internal Imperatives

Wole Soyinka

Every thinking inhabitant of a given national space must surely, at some moment or other, reflect upon the significance or insignificance of his or her own identity as it relates to the existing, or historical, definitions of that space. This questioning may have a simple focus, such as language, and may be taken no further once that issue is, apparently, satisfactorily resolved: the Welsh, for instance, or the Occitans in France, the Berbers in Algeria, or indeed the Hispanics in the USA. It might also be manifested in a wider cultural awareness — a restless sense of identity that might dangerously stress the casing of a common nationality. Economic arrangements — usually the unsatisfactory aspects, as they are perceived by different groupings within the overall national body — may also cause such questioning. Expressed in the language of 'unequal development', 'unfair revenue allocation' or 'regional neglect', this last factor is invariably a triggering mechanism for the recollection of some other identity, one that has been subsumed, however briefly or historically dated, beneath existing assumptions or impositions of nation-being. The Soviet Union provides the most chastening instance of this.

Less known, at the moment, is the tragic example of the Ogoni people, victims of a genocidal onslaught by a singularly vicious military dictatorship in my own Nigeria. The Ogoni predicament has provoked, sometimes in the most unexpected quarters, this exercise in national introspection. Here I refer to open debates which increasingly test the assumptions of nation-being — whether as an ideal, a national bonding, a provider, a haven of security and order, or as an enterprise of productive co-existence — against the direct experience of the actual human beings who make up the nation. This is when we are confronted with the ground rule that any perception of the nation which is rooted in ideas not shared by the human beings who are the fundamental elements of its very existence, is as vaporous as the nation itself. We might ask ourselves: why do the citizens of the USA periodically set upon their flag, trample on it and burn it? What makes the phenomenon so prevalent as to require legislation that criminalizes this act of rejection and desecration? Why is there, today, an extremist tendency that accuses the state of becoming a structure of alienation and has even resulted in the formation of local militia, whose members have taken up arms to defend their nation-definition?

Sadly, this truism is not as universal as it deserves to be; or at best, it is one that rarely transcends lip-service. In truth, some conceive of a nation almost exclusively in terms of Gross National Product. Others — a disappearing breed — see nation-being as a material manifestation of a remorseless, historical process which they are singularly privileged to preconceive and direct. Still others, with increasing virulence in recent times, conceive of the nation as an expression of Divine Will whose active processes, from private conduct to the arts, from fiscal policies to architectural designs, must be governed by the desire to win approbation of, or reflect and glorify the omnipotence and grandeur of, the Invisible Presence. To many, the 'City of God' is not a metaphor, unless it is a mere diminutive for the 'Nation of God'. Even as I speak, the number of journalists murdered in Algeria in the name of a Supreme Deity and his apotheosized prophet may have reached fifty. (The figure last stood at forty-seven, but that was all of a week or two ago.)

If only the sacrifices demanded of the human polity by those with such extra-terrestrial perceptions of society were limited to the non-consultation of that humanity, even by proxy! If only the affected had nothing worse to complain of than the failure of such a divinely endowed elite to make the rest an active part — not a coerced, cowed and submissive pawn in what are, after all, undertakings within and on behalf of the total community — one might be able to shrug off the questionable priorities that result from such arrogation of power and resources by the few on behalf of the entirety! But people often suffer consequences more serious than a mere emotive response springing from the sense of being ignored. However, let us not burden ourselves, for the moment, with the statistics of murders that are being planned or executed, at this moment, in a country like Egypt or Algeria, in the name of the Divine Will. We shall attempt instead an easier task, restricting ourselves to the theme of development, and the politics of development, asking whether or not we can identify any stronger criteria for addressing a populated space as a nation.

Let us take the example of a distressed environment, impoverished, slum-ridden: against all the laws of organic development, it is made to undergo the intrusion of a dominating, environmentally disproportionate structure.

Perhaps even this could be endured: if the existence of such a structure had no deleterious social consequences, from conception through to realization, one could console oneself with the resultant monumental landmarks, some of them indeed of enduring and aesthetic properties. But we do know that these heaven-directed labours of love are achieved at the expense of the fundamental well-being of the human entities below. The replication of St Peter's basilica in remote Yamasoukrou, birthplace of the late Ivoirian President; or the no less extravagant 'triumph of modern architectural engineering' (one of the more restrained descriptions) of the mosque in Casablanca — are these really expressions of devotion to the unseen? Or are they delusions of grandeur, the craving for immortalization by some individuals who find themselves in, or who manipulate themselves into, positions where, without the responsibility for statuary accounting, the entire treasury of a nation is placed at their personal disposal?

The question is whether private whims, even of the deepest religious colouring, are really a rational substitute for the systematic identification of development priorities for a nation. All we can be certain of, because it is clearly provable, is that the proliferation of grandiose cathedrals, basilicas, temples, mosques and shrines throughout the global landscape has not perceptibly improved the living conditions or moral sensibilities of the large part of humanity — judging either by the physical conditions of the populace where these architectural caryatids are situated, or by their social conduct. Jean Genet was so perceptive: what are these structures but private mausoleums, paeans to human vanity at the expense of social realities?

So the question remains: What is the justifying rationale, viewed from the perspective of the millions of beggars who litter the streets of those countries, the hit-and-miss level of their health delivery services, or indeed the inability to provide their teeming generations with minimal access to the most basic structures of self-fulfilment in a modern, competitive world? We know that the foreign architectural consortia benefit. They hone their skills on these engineering challenges and swell their capital reserves by billions of whatever currency they elect to be paid. Their skilled workforces are guaranteed a livelihood for years. But the nationals themselves? Apart from unskilled labour, just what do they productively contribute? What level of integrated development does the nation gain? How interactive are such structures in the enhancement of the quality of daily life?

* * * * *

It is against the background of such travesties that one grabs deliriously at the first paragraph of the preamble of a United Nations document, which is none other than the final presentation by African Ministers, responsible for Human Development, to the March 1995 Copenhagen Conference. This paragraph goes to the heart of my contention and relieves me, and others in my position, of the need to assume burdens that really belong to others. It is a curious statement altogether. Obvious truths — especially when they carry the flavour or implication of an accusation, of a dereliction of responsibilities, a misdirection of social endeavours — tend to arouse combative responses in governments when they come from writers or intellectuals. It is a wearisomely familiar self-defensive reflex, one which manifests itself in those demonizing expressions: 'utopian', 'unrealistic', 'cloud cuckoo-land' — even that catch-all dismissive taunt, 'obviously Westernized thinking'.

Those who most frequently deal in such language (usually spokesmen of African governments and a handful of establishment intellectuals, the super-patriots) little realize that they insult history and discredit their own race by the attribution of humanistic principles solely to the Western world. We are accustomed to those knowing, even patronizing winks and nods that supposedly imbue such agents of government policies with the benefit of inside knowledge, a superior foundation of experience and so forth, all of which guarantee that the implicit accusations or questions are never answered, only deflected, and the propositions evaded as something only to be expected from those 'alienated minds'. Yet it appears that a reformation in thinking has been taking place in those very uncontaminated minds in recent times! Is it all convenient rhetoric, an attempt to sanitize the disreputable power images that they represent before those far more potent powers from whom aid or co-operation is solicited? We shall insist on taking their statements at face value.

So, with no claim to originality, and employing the very words of bureaucrats, technocrats and politicians, the following must be assumed to be the coincidence of their views on the human element within any concept of development:

> We, the Governments of African countries represented by our ministers responsible for human and social development, meeting in Addis Ababa on 20 and 21 January 1994 at a preparatory regional conference on the World Summit for Social Development to be held in Copenhagen, Denmark, during 6 to 12 March 1995, resolutely affirm the centrality of the human being as the initiator and beneficiary of development, the means and the end, the agent through whom and for whom all development activities must be undertaken.

What is new here? Nothing, except of course the source from which this declaration has emerged: 'We, the Governments of African countries'. Since we may never hear it again, let us celebrate this welcome aberration, hoping that it has not cost many of its formulators the sack or worse.

This declaration, in my view, represents a final convergence of nation-definition and fundamental expectations between the rulers and the ruled, between the leaders and the led. For what else have the latter ever claimed? On what other platform have they ever agitated? For what other cause have they suffered imprisonment, torture, exile, even death? Examine that summative declaration of faith how you will, through whatever ideological prism or with whatever sententious hair-splitting, and from whatever terrain — from former apartheid South Africa, through former communist ortho-doxies to the present scourge of fundamentalist insanity that now threatens to coerce the world into its own demonic closure of nation-being. The essence of that declaration has remained one constant factor, one bedrock against which every creed, every ideology, every racist philosophy, every cult of power has been hurled again and again in tidal waves. The bedrock has not budged, and its adherents have clung to it like limpets, defying the direction-changing storms of ideological imperatives. Let us not forget, however, that millions have drowned in those storms: millions have been swept away, have perished, never really understanding why, never really understanding to what gods they were sacrificed, knowing only that the state, or the aspiring state, had ordered it; that some programme in the cause of a mere concept of nationhood demanded they be uprooted from their homes, turned into stateless non-persons, degraded from creatures of feeling or sentience to mere digits in some abstract evocation that had become an end in itself, not the means to the elevation of humanity, the enhancement of its productive potential or the harmonization of its relationship with power and authority.

We have been compelled to live decades and centuries of lies, often propagated with state violence — lies that are often compelling, since humanity, to its credit, is never content with the limitations of its own material body and immaterial spirit, and therefore seeks goals that extend beyond itself, beyond its immediate seizure of, and relationship with, the world. It is a human characteristic; and myriad, we acknowledge, are the attainments that owe their genesis to this restlessness of the human mind. Alas, there are those individuals who are especially gifted with the exploitation of this craving for human self-enlargement. They are the con-men of society, experienced liars, solipsistic manipulators. They belong to the same tribe of smooth, practised salesmen who can offload the proverbial shipload of fur-lined polar boots on a tribe of tropical herdsmen and throw in a consignment of snow tyres for good measure. That analogy, I promise you, is not far-fetched but borrowed from reality. Many notions of nation building and development on our continent have proved as relevant to actualities as polar boots on the feet of a Masai herdsman.

Some of these lies are still with us, especially in the territory of the ineffable, called religion. Spaces that were once teeming human habitations are now depopulated because of these compelling projections that over-whelm the material rootedness of millions and destroy their well-being. For instance (to return to the physical zone of existence), wherever the nation-ideal becomes a notion beyond the centrality of human livelihood, wherever the nation-ideal becomes conflated with notions of racial purity or other forms of extreme nationalism, we know only too well what the consequences have been. We encounter exclusivist policies that go beyond expelling other human units beyond a specific national space. The new imperative demands that they be excluded from the category of humanity, and thus from the physical world altogether. Ideological rigidity, religious extremism, racism or xenophobia — these are all foundations of forms of transcendentalism that have dire consequences for any people.

We are familiar with all the predictable rationalizing, mythification processes. We know the demagoguery that overcomes the susceptible and makes millions submissive to the hypnotic mechanisms of power. From time to time, however, a portion of that same population of victims recovers its senses, begins to function again as part of a human collective that has become conscious of having been robbed of its will and is ready to contest the robbery and demand restitution. My understanding, my guarded assessment, of the grudging moves towards the restoration of humanity (rather than abstractions) at the centre of development, as demonstrated in the cited declaration of the representatives of African governments (maybe we should frame it and hang it up in all public buildings, hospitals and schools) is that perhaps, just perhaps, the mutual admiration society of African leaders is preparing to join the rational world in repudiating this wasteful cycle of loss and recovery, and has chosen to begin at the only starting and finishing point of development that can be universally agreed, because it is not a projection of the imagination but a material, irreducible factor — the human entity.

*****

In addressing the issues of nationhood and development, therefore, both in definition and as a living project, the immediate question with which we are confronted is: How is the collectivity of such a unit best organized? Or to begin with what we know, what we observe, live in, die in and even die for: How do such national entities currently fare, those present groupings of that lowest common denominator, the human unit? Are they working? Or do they work against the constituent units, the human beings? In short, does the superimposed idea ('nation') harmonize or conflict with our given a priori ('humanity')? If there is conflict, what are the causes? What are the histories of nations, and what future do such histories further threaten? Some of these questions answer themselves very quickly and with a chastening finality: Rwanda for one; or Yugoslavia; or the Sudan. Others permit the luxury of hesitation, the lingering over uncertainties, the weighing of pros and cons, some space for tinkering and time for the claims of redeeming factors. But clearly, the responses in the case of a few so-called nations are stark, unambiguous and clamorous, even desperate for remedial action. It is from that human perspective that any national surely pauses periodically and demands: is mine the ideal state of the nation-idea? A question, or stock-taking, which can only be addressed on the material plane, by looking at the ledger-book, not at mysterious texts of pious intentions or abstract notions. Humanity is not abstract.

We have spoken of a competitive GNP. Those who think that this alienated concept of humanity was limited, in Africa, to forced labour and distorted production systems, most notoriously under Belgian colonialism (or, further afield, to the dehumanized production collectives of the Soviet Union) had better take a second look at the economic experiments — first of countries like Tanzania and then, to a far more destructive, indeed criminal degree, the forced human displacement programmes of Mengistu's Ethiopia, in the name of ill-digested notions of Marxist economic centralism. This is not to suggest that such impetus in nation building, or transformation, is always so idealistically motivated (if one may be permitted to abuse that expression in a context which involves, sometimes, the decimation of the very people who contribute to the attainment of such nation-ideals). We need only recall the fate of millions of kulaks under Stalin, or the annual trail of skeletons that became the identifying mark of Mengistu's economic policies.

Moving away from the categories of the 'idealist' builders — however warped those ideals may have proven — there are others at the comic end of the tragi-comic axis, whose recollection of nationhood is triggered only during Independence Day or other calendar occasions, and during sporting events, especially football. Perhaps I should have said farcical: comic is too generous a word to describe what strikes me as a purely jingoistic invocation of one's nation in those spasmodic recollection habits which exclude the apprehension of the nation as a continuous living organism, as one that shares the same basic human component as other, near or distant but evidently productive claimants to that definition. It is difficult to recall, for instance, that a war was actually fought for the sake of football. The nations involved were, I believe, Honduras and El Salvador. This supplies us with the quintessence of farce as a promiscuous begetter of tragedy. When one reflects on this normally comatose nationalist agenda, one that is virtually non-existent in the mind of the ordinary toiling peasant or worker, but for which he is called upon to kill and be killed at the whim of a dictator, the words of the Anglo-Irish poet and playright, William Butler Yeats, come readily to mind: 'State and nation are the work of intellect, and when you consider what comes before and after them, they are not worth the blade of grass God gives for the nest of the linnet'.

Sobering words, for the recollection of which I must thank Denis Healey in a recent article in the London *Sunday Times*. Also for the reminder that Yeats was a fierce nationalist in his early writings, who came, towards the end of his life, to recognize nationalism as 'a dangerous illusion'. It can of course also be a farcical illusion, an opportunist, purely adventurist evocation, with tragic overtones, which is where we are at the moment, and which compels us to acknowledge the impassable gulf in nation-apprehension between African forms of leadership, the dictatorships most notoriously, and the populace itself.

Moving down from divine architecture to mundane, bruising preoccupations: how can one explain why the ruler of an impoverished, bankrupt state would seek to deplete that nation's resources even further, for the sole purpose of staging a World Cup event on its tortured soil? It can only be explained by the need for national self assertion — a somnambulist response to a decaying reality, where the undertaker of a moribund nation suddenly wakes up to the existence of an entity he hardly comprehends, except that it shares a category called nationhood with others and is therefore entitled to contest certain functions with them, in this instance, sports. For fear the point may be missed, let me belabour it without apologies: in the conception of such a ruler, a Junior World Cup final, on his own soil, presided over by his person, is the equivalent of the Basilica of Yamasoukrou or the Grand Mosque of Casablanca. It is a triumphal arch that magnifies his puny being, a perpetual mausoleum within a legend that monumentalizes his otherwise unremarkable passage.

The African nation, alas, is mostly viewed through the goggles of such rulership, in studied contrast to the far more organic apprehension of that word when applied to entities like France, Sweden, Japan, Italy, South Korea — most of the European nations, a few Asian and American. It is the confidence of such national entities, their ability to take their status for granted, that makes it possible for them to embark upon new arrangements such as the European Union (with all its attendant bickering and retreats), to argue over serious and petty details, even to be 'nationalist', but in a concrete sense that goes beyond the slogan of defending 'national sovereignty' and protecting 'our way of life', to protecting trade, fishing rights, jobs, labour standards, or contrasting their national realities in a positive manner with those of other member states. What, by contrast, has a geographical space such as Sierra Leone or Gabon or Nigeria come to mean, in concrete terms, when a leader speaks of 'national sovereignty'? Does that expression have the slightest impact on the market woman or the factory worker who sees the buying power of the nation's currency dwindle to nothing, who sees health as a luxury reserved strictly for the affluent and exhibitionist cronies of power, and social opportunity the hereditary preserve of their pampered scions? Is it anything more than an echo, void of substance once it has left the mouth of an incumbent dictator? A nation is a collective enterprise; outside of that, it is mostly a gambling space for the opportunism and adventurism of power.

*****

Now let us ask ourselves which of these perspectives represents the Nigeria that we know today. What is the social reality as experienced by the inhabitants of that space into which a footballing basilica was about to be intruded? Is it by any chance free of the population of beggars that crowd Casablanca or Yamasoukrou? Of vast sections of disease-ridden humanity? Are the faces we encounter radiant with hope and confidence in their future, or masked with fear and uncertainty?

The response is stark and unambiguous; but let us begin by recollecting that it was not always so, that fortunately there are those who recall a social cohabitation that was not riven by the present uncertainties. There were times when religion was a harmonizing factor even between communities of different faiths, when a spiritual richness pervaded daily existence no matter whether it came from the Moslem, traditional or Christian social and religious structures and observances. The use to which religion is put today — and we speak here not merely of extremists but of government complicity — often translates directly into politics, both local and national. The nature of such politics does not require much effort to envisage.

There have been too many lost moments, moments when this particular disease could have been firmly rooted out, when leadership chose instead to exacerbate such divisions for its own agenda of control rather than to set an example in the harmonization of faiths. We are speaking — to name a concrete instance — of a nation of multiple faiths that has yet to recover from the effects of being dragged into corporate, national membership in the Organization of Islamic Conference. It is not a question of whether or not membership in such a council would be advantageous in some form or the other to sections of the nation or even the nation as a whole — aid and soft loans from Islamic states, commercial privileges and so forth — but simply that this was one situation, one mightily explosive distraction, that the nation could have done without. It was a move that was bound to divide the nation and exacerbate sectarian suspicions, which of course it did. Religiosity has very little to do with nation-to-nation embrace of whatever kind, and even less to do with those that carry the slightest suspicion of political overtones. The reverberations of such a move — what we might call the ripple effects — continue to plague Nigeria today, encouraging a disruptive militancy in some and a disruptive counter response in the rest. However, let us not remain too long in the spiritual realm. The agony of a nation is observable largely in material reality, which alone records its proof of existence. How is that reality experienced by the Nigerian people on a daily basis? Let us begin with the forum for the exchange and dissemination of ideas and social awareness, one that at least guarantees a measure of participation in the flow of life within any community.

The press, that once vibrant voice of the Nigerian nation, is officially dead, wiped out in one fell swoop, unprecedented in the history of Nigeria. Apart from the valiant remnant of that press, functioning with an erratic sword grazing its neck, the nation's voice is reduced to two newspapers run by the government — one of them bankrupted by lack of readership — as well as the electronic media, across which that dictatorship has placed a stranglehold. The once stimulating, provocative discourse of a nation of at least ninety million people is reduced to mendacious bleats from the seat of government, and the relay system of a sycophantic train. But the existence of a vigorous, risk-taking underground press speaks volumes; it remains one of the signs that the combative spirit of a nation is not yet extinguished.

The health services of that nation are non-existent; mothers die in childbirth for lack of the most basic drugs and a hygienic environment for labour. Infant mortality has reached alarming proportions. The simplest, most easily curable diseases flourish for lack of treatment — and kill. Three years ago, a military governor — a pillar of the establishment — lamented that the hospitals had become mere consulting clinics, so desperate was the dearth of basic drugs and hospital equipment. Two years later a medical specialist, also in the service of government, was compelled to return to that statement, qualifying it with the comment that the nation's hospitals could no longer claim even that dubious status, but had become virtual mortuaries.

Potable water, even in the heart of the nation's capital, has become a commodity that is left to the dispensation of the skies, from where, perhaps, it is also hoped that Sango, the god of lightning, might perform the miracle that has so far eluded the Nigerian Electric Power Authority, an infamous institution known as NEPA. (Nigerians have become bored with the game of finding new readings for that acronym, the best known being 'Never Electric Power Anytime'.) Many small businesses have collapsed as a result of this failure, and join domestic households in mourning the periodic vengeful surges of power that incinerate their appliances and even their homes, as if the god of lightning has indeed taken personal charge, but without any schooling in the elementary laws of voltage distribution. The generator trade — not industry! — booms regardless, the market monopolized at the top by the very public servants who are paid from the public purse for the supply of electric power.

Public transportation is so inadequate that it provides a study in collective masochism, often degenerating, at the arrival of the lone tumbrel, into a contest to determine the survival of the fittest. In certain suburban areas such as Oshodi, Apapa and Ipaja, workers on a seven-thirty to five o'clock shift must leave home at four in the morning and are lucky to be home at eight or nine o'clock in the evening. The sight of in-between commuters listlessly awaiting some form of locomotion on a blistering Lagos afternoon, at improvised stops without shelter, while the latest off-the-line models in private motoring cruise by, strikes visitors no less forcibly than it hits those who still guard their conscience among the privileged of such a national space. It is over twenty years since I wrote *Opera Wonyosi*, and while the social condition of the people had not yet deteriorated to the present level of despair, this (easily manageable) condition was already so aggravating that I accorded it a verse in the play:

> Have you seen those workers daily jostling
> catch a bus to beat the factory deadline?
> And the pregnant mother wedged with elbows
> Barely dodging those haphazard blows
> You'll claim the boss is also on the breadline
> The 'go-slow' has wrecked his daily hustling
> Well, a whole day in an air-conditioned car
> is sweeter than one hour in over-heated air.
> Chorus: Explain the smugness on the face of the chauffeur?
> He knows that at the bus stop life is even rougher.

Today, a different kind of language would be required to address this violation of the human essence! And what of education? Tertiary institutions have fallen below the level of secondary schools (let us take note of this; it is a statement that will be echoed later within that same declaration of African government spokesmen), while hundreds of thousands of youths roam the streets, jobless, without purpose or direction, half-baked products of secondary and tertiary institutions.

Hunger stalks the streets and, with it, desperation. Thus the security of individuals has become a game of Russian roulette — one never knows whose turn it is. Even diplomats have been compelled to lead protest delegations to the seat of government in Abuja as they regularly discover the limitations of diplomatic immunity, be it on the open road, in the heart of the city or in their residences. It is small consolation that — unlike journalists, writers and women in Algeria under fundamentalist malevolence — the average Nigerian citizen living in the larger cities or compelled to travel the open road, has not begun to write his obituary in advance. (Still, some do dictate, with a fatalistic effort at gallows humour, their last will and testament before leaving home.) Yet the country that serves us here as a model is poor neither in human nor in material resources. Only last year, a government-appointed commission of enquiry concluded that over $12 billion, the windfall in oil revenue from the recent Gulf War, were missing, unaccounted for. Millions more continue to vanish every day, in magical disappearing acts or in misdirected ventures.

This is the daily reality of the environment from which a dictator and his yes-men, without a shred of embarrassment, presumed to launch the rags of national identity into the international arena. From a comatose habituation, those who claim to articulate the national will are suddenly aroused to the potential privileges of a living organism. Like long-deprived drug addicts injected with their favourite poison, they are jerked into consciousness and realize — oh yes, of course, we are a nation, and we have national rights. We must assert ourselves as a nation. And we must put the nation on the world map — oh, that mangy, flea-infested flag that the sanctimonious nationalist drapes around his torso to cover a repulsive nudity!

All those who disagree are subversives and traitors and lackeys of foreign governments. The world tournament stays here, even if we bankrupt the nation in the process. It entails further degradation of the quality of existence for our people, but so be it. Somebody must pay for our moment of national accreditation. Not so, said the people. There is no nation within this space to host the Junior World Cup Championships, not in 1995. And they took to the streets, and three days of fasting were ordered by church and Mosque leaders within the so-called national borders, and the band of 'traitors', 'national saboteurs' and 'unpatriotic elements' swelled both at home and abroad, until the footballing club of nations began to take note and was compelled to relocate its fiesta. From Nigeria to tiny Qatar — quite a comedown for the self-vaunting giant of Africa!

*****

I stated earlier that it was not always so. It is good to remember this. So how did it all go so badly wrong? I am bound to prolong this focus on my immediate geo-political constituency of Nigeria, recalling, however, that for many of my generation, that constituency was not always so restricted, but was indeed a continental state of self-identification. I know that as we came into self-awareness as productive beings, we brought our immediate national space into perspective, not narrowly as an idea from which we took a socio-political definition of ourselves, but as a branch of an even larger idea — the idea of a continental identity in formation. Such nationalist vision as we had transcended our own boundaries. Many students of my generation surely set their political sights on variants of a continental oneness: the colonial settler regimes of East Africa, and Apartheid South Africa in particular, dictated this racial challenge. We were destined — this much appeared so gloriously clear to us — to be at the forefront of Africa's version of the International Brigade. Our liberation hordes would sweep down from Ghana, Nigeria, the Cameroons to engulf the colonial settler regimes of Kenya, the Rhodesias, stopping only at the most provocative tip of the continent — South Africa — whose apartheid philosophy and policies had moved beyond the terrain of experimentation to the most extreme, hallucinatory edges of dehumanization. (What colour are the hands that dehumanize our African peoples today, as they have done for nearly four decades of independence?)

In between, the forests of Kenya were already sutured with the elusive motions of the Mau-Mau liberation forces, criss-crossed by British colonial hounds in desperate pursuit. Where these motions intersected in time, the result was death or misery in British detention camps. Heirs to various pan-African movements (beginning with the famous Manchester gathering of 1945), Nigerian, Ghanaian and East African students debated news from these and other stubborn enclaves, viewing them as assaults on both racial and national pride. So deep was this sense of total repossession that our nationalists, in Nigeria at least, sang the praises of the mosquito, crediting that malaria-dispensing scourge with the failure of the European colonizer to occupy the West coast with the same settler lust as was inflicted on East and South Africa. Let me sum up this vision of the nationalists, a vision that was to be fulfilled by the liberation forces of which we were destined to be the vanguard: we saw the continent, at least from the south of the Sahara to the southern tip of the continent, not as a conglomeration of nations, but as one nation, one people.

For that same Nigerian, however — and this is certainly true also of the Ghanaian, the Senegalese (with Arab/Mauritanian complications), the Malian, Kenyan, Malawian, Zairois — the boundaries of a communal identity are today set much more narrowly. The visions of the average nationalist have sadly contracted. The reason is simple: there is so much work to do; and charity, it is said, begins at home. Most Nigerians of the pan-Africanist temper have moved away from the continental shelf to the boundaries of colonial endowment, desperate for a salvage operation of what is closest at hand. In some cases — and here we come to the real predicament — the national capsule is even seen to have cracked internally and the watchers have begun a drastic interrogation of history, of the beginnings, and of the affective meaning of the national identity. It is not an elevating development, and it is one which many, in well-insulated compartments of governance, prefer to pretend does not exist. But it is real! Despite the pains of any official ostrich, it is only too real; and, for us in Nigeria, it commenced over the past decade and has dominated public discourse most critically in the past two or three years, to situate the phenomenon in its precise, immediate cause — the annulment of the elections of June 1993. Let us hasten to absolve the Nigerian populace, the ruled, almost in its entirety, from this regression into narrowed entities. We must identify the cause where it manifestly is, where it is always to be found: in a minority that constantly plays up innately innocuous differences, be they of ethnicity or religion, in order to set one section against another and thus assure itself of political control.

It is customary to declare that man is a political animal. It is a definition to which I subscribe, but for purely strategic reasons. We must insist that man is indeed a political animal if only to give dictators sleepless nights, in order to remind them that they hold sway over a restless breed of their own kind, whom they have deprived of their animal rights, and who will one day challenge them, true to type, over this territorial imperative, as usage dictates within the animal kingdom. When, as the fable teaches us, the king of the forest loses all measure of self and demands that his daily prey deliver itself up to him in meek obedience to an agreed roster of consumption, it is time to confront him with the fathomless resource of alternative power and drown him in the arrogance of a delusion.

In a more fundamental sense, however, man is first a cultural being. Before politics, there was clearly culture. Only man the producer could have evolved into the political being, which, to pare away all mystification, is the evolutionary stage related to the development of society and the consequent sophistication acquired in the management and protection of resources. This hierarchy of evolution also explains why man resorts to his cultural affiliations when politics appear to have failed him, never the other way round. The ongoing retreats into real and pseudo-ethnic bonds on the Nigerian political scene, which can also be witnessed in other evolving societies both in Europe (the former communist empire especially) and on the African continent, can best be understood in this light. With this reminder, we again address the rock on which the Nigerian ship appears set to founder: the democratic leap of June 1993.

Any internally robust, well consolidated capsule can withstand a certain level of stress from within or buffeting from without. For an already distressed enclosure, however (such as we have taken pains to depict as accurately as can be universally attested) — one whose inmates nonetheless remain conscious of their own potential — a simple test prod of its casing (again, from within or without) can prove to be more than the mere letting out of noxious gases; it may prove the final, fatal prod, resulting in the quiet collapse of the balloon or the implosive rupture of a deceptively hermetic edifice. 'In my father's house, there are many mansions', runs the spiritual. When any one mansion is prodded towards a recollection of decades of wrongs, but more crucially, towards a feeling of a conspiracy by elements within another mansion, a minority whose vision of society is an expansive condescending patronage, one whose relationship even to the inmates of its own specific mansion is that of the hereditary overlord in a tradition of serfdom — then need we be surprised when every inmate becomes a Structural Inspector, taps on the walls and reports: 'Unsound. De-certified for human habitation'.

Go to the markets, go to the mechanics' villages, mingle among the 'Area Boys' of Lagos and Kano, travel incognito in a long distance bus from Agege to Benin, Okene, Abuja, Kaduna, Sokoto, Maiduguri, speak to these 'unlettered' inmates of unprivileged mansions of 'my father's house', and the object of their rancour is inescapable: one mansion (and not even its entirety, but a chamber — the most luxurious, predictably), the occupants of which have developed a chronic propensity for alliances with kin interests from other privileged habitations of the total household. And the life-style and life mission of these indolent, spoilt scions of the household render insecure the foundations of a simple enterprise of cohabitation. Inevitably, other dwellers come to the question: is it not more sensible to pull the rug from under such pampered feet by establishing our own self-subsisting habitation?

So, if you pose the question to such sections of that nation today, in the form 'Do you believe in Nigeria as a nation?', the answer from many sections, going by the tenor of debate from 23 June 1993 until now — be it within the 'mansions' described above or, to transfer the discussion to a formalized 'representative' forum, within the diversionary, spendthrift, so-called Constitutional Conference (a collective insult that was imposed on the nation by the dictator to ensure an eternity in office) — the response we obtain is weighted on the side of 'No!'. 23 June 1993 was, of course, the date when the nation was stunned with the annulment of the presidential election of 12 June, an election whose results were already halfway broadcast to the whole world, and whose conduct was universally adjudged free and fair by any standards. The perpetrator of this 'crime against Nigerian humanity' was none other than the former major-domo of that feudal mansion that is resolved to impose its anachronistic and undisciplined vision of house-keeping on the rest of the family compound.

At best — to return to our question — we might receive an ambiguous, qualified 'Yes', hedged about with conditions; but those who wish to be truthful to the evidence must admit that the mood of the nation, in the main, is inclined toward a vote for the reconsideration of the nation-status as it now exists. The irony is that it was not the Nigerian populace who repudiated nationhood. On the contrary, they expressed, on that day of

12 June, a clear desire for nothing else. It was the numerically infinitesimal but well-positioned minority, blinded by self-interest, seeing that a nation was about to slip from its hands and be restored to a majority dispensation, that commenced the destruction of all sense of belonging. It is that same minuscule proportion which, having succeeded in robbing the Nigerian people of their nationhood — at least for now — insists, with the dismally predictable collaboration of their foreign, traditional (colonial) backers, and in face of stubborn daily evidence to the contrary, that the famed election of 12 June 1993 is ancient history. Alas, if that election proves indeed to be ancient history, then — and do take this as prophecy — Nigeria as a nation has no future history.

The perpetrators of that deed, and their apologists, plead a pragmatism which only they appear to understand. The mandate given to Nigeria's President-elect, Basorun M.K.O. Abiola, they insist, has become tainted with the passage of time. It has become compromised beyond salvage by events to which, they claim, that candidate was himself a party. Maybe; maybe not. The fundamental issue (conveniently ignored) is that the elections of 12 June 1993 are totally beyond the control or directives of the principals. The apologists of the annulment of that election sadly do not understand where their logic leads. If the will of a nation, freely expressed, attested to by observers from the international community as an exemplary democratic exercise, endorsed even by the monitoring agents appointed by the incumbent regime and by its secret services, if that expression of nation-being, since it was a result that cut across all boundaries — ethnic, religious, professional, class, and even across the established parties — if such a ringing declaration of a hunger for nationhood, of a craving for a democratic order is so tainted as to become invalid, what does that make of the nation whose will is so easily flouted? Any election — leaving aside for now the history of this specific one — can be either the cement that binds a nation together or else the porous vessel through which its life-blood seeps away. If the nation's will has become so tainted that it cannot be implemented, then the nation itself has become so contaminated that it cannot begin to claim the recognition of a nation.

*****

When I find myself cornered by others who pose a similar question to me — 'Do you believe in Nigeria as a nation?' — my answer is invariably a question in response: 'Do you accept my definition of a nation?' Now I find my definition endorsed by the declaration of representatives of African governments, lodged in the documents now filed with the United Nations. Thus, for the moment, I am able to claim that I accept Nigeria as a duty, that is all. I accept Nigeria as a responsibility, without sentiment. I accept that entity, Nigeria, as a space into which I happen to have been born, and therefore a space within which I am bound to collaborate with fellow occupants in the pursuit of justice and ethical life, to establish a guaranteed access for all to the resources it produces, and to thwart every tendency in any group to act against that determined common denominator of a national social existence. I accept that space as a space of opportunities and responsibilities that must extend beyond its boundaries, principally because of its rich endowment in material resources. I accept that space as one that is best kept intact, in order to harness those resources with maximum efficiency, to conserve and mutually cross-pollinate its cultural hoards, and to enable it to link hands with others right across to the southern tip of the continent and present a formidable machinery of collaboration on an equal relationship with the rest of the world.

Expressions such as 'territorial integrity' and the 'sacrosanctity of boundaries' — those relics of a colonial master-slave bequest that abjectly glorify the *dictat* of colonial powers — are meaningless in such a context. The supporters of geography as an instrument of patriotism belong, in my view, to the class of people referred to by that writer who declared that patriotism is the last refuge of the scoundrel. What is in it for those within that cordon of geographical patriotism — that, clearly, is the ultimate question and purpose. If one accepts Nigeria as a space which must move beyond what a politician once described as a 'mere geographical expression' to what my vision dictates as a humanized space of organic development, then I may be convinced to stop quibbling over mere nomenclatures. Until then, that unfulfilled promise, Nigeria, must remain only a duty. It is that same duty that we, on our part, must continue to urge upon those same 'Governments of African countries', challenging them to realize their own pronouncements, denouncing them before the entire world when they fail to do so, and insisting in such a case that they be treated as pariahs, as the real traitors, to their own kind and to humanity in general.

Clearly, that space, Nigeria, cannot be the duty and the burden of the writer and the intellectual alone. Indeed, our function is primarily to project those voices which, despite massive repression, continue to place their governments on notice. Additionally, however, we insist that it is time to move from quiet diplomacy to variants of the same concerted action that brought to its knees one other nation, apartheid South Africa, which refused to accept the universal affirmation that makes the centrality of the human being, without exception of colour, class, race or religion, the means and the end, the agent through whom, and for whom, all development activities must be undertaken.

* * * * *

That declaration, which I persist in attributing to its unlikely source, the representatives of African governments, summarizes the predicament of our society. It recognizes a definite crisis of nation-being and furthermore provides unambiguous details, with brutal assessments of numerous aspects of national life throughout the continent. It identifies, for instance, 'a crisis of governance encompassing such well-known shortcomings as the near absence of democratic structures, popular participation, political account- ability and transparency...'. It recognizes that 'civil strife is closely associated with challenges to authoritarian structures of government, as well as ethnic and communal confrontations'.

These government spokesmen could have added there, without contra- diction, that such communal confrontations are often deliberately provoked into being by authoritarian regimes in order to create instability, which then justifies repression and an excuse to remain longer in power until 'order is restored', 'the shattered economy recovers', 'a new constitution is written and adopted', 'stability is guaranteed', 'the crisis' — which they fomented — 'is resolved', and so forth. However, our guiding representatives of governments do continue to agree with us — the alienated, nation denigrat- ing, brainwashed Westerners, negativists and professional doomsayers, whose sense of sophistication is simply to pull down their own peoples before Western eyes. They join us in observing that 'over twenty million Africans are refugees and displaced persons. This represents almost half of the world's refugees'.

This statistical ignominy appears to be even beyond our own remedy, for within those boundaries, the so-called 'nations' that at least organize their own existence (and upon whose shoulders the responsibility for the relief of their own kind ought to fall) appear indeed unable to salvage their own internal predicaments; how, then, can they look beyond their borders to ameliorate the calamity of the flotsam and jetsam of humanity in makeshift refugee camps! No, they must await the advent of the next redeemer in the shape of a Bob Geldof or the United Nations. The stark reality that cripples rescue efforts from within the continent is acknowledged in the same report:

> concomitant with the decline in all the indicators of human and social development is the virtual collapse of African institutional capacity. Hospitals and health centres lack basic equipment and amenities; schools lack basic teaching aids and necessities such as chalk; African universities and institutions, once the training ground for the region's leaders, professionals and technicians have now become poor performing institutions. The morale of those working in many of these institutions has reached rock bottom ...

Now, surely, as I hinted earlier, the litany of 'those dissident voices' and that of the government representatives have become really inseparable, one from the other. And thus it goes on and on, the dismal picture of a continent that continues to smile at its image in the mirror while the whole world looks on in tears: 'the need to accommodate freedom of opinions, tolerate differences', 'the need for full and genuine participation in the political, economic and social processes of their countries', 'the positive side of a revival of religious practices and beliefs, *in so far as tolerance can be inculcated* as a crucial dimension of religious culture' (my emphasis). That dimension has of course become a matter of life and death for the individual, for groups, for women especially, and consequently a matter of life and death for the nation itself. Let us wind up the list of indictments handed down by these government ministers and bureaucrats with the one that concerns us most today: the ignoble and retrogressive role of the military in the African crisis, their wasteful, unaccounted spending, their corruption, their alienated apprehension of society and nationhood, and their brutal repression of civic aspirations.

Here, at least, we may be able to reassure these official critics, who may feel uncomfortable about the kind of company in which they now find themselves, that they can find far more respectable — and still African — support in the office of the Secretary-General of the Commonwealth of Nations, now held by Chief Emeka Anyaoku. In his address in Abuja, Nigeria, on the occasion of the first Memorial Lecture in honour of the first elected head of a Nigerian government, the late Sir Abubakar Tafawa Balewa, Chief Anyaoku brandished a scoresheet of catastrophic grades for military regimes as one long record of incompetence and inbuilt instability. The function of the military, he emphasized, is to defend the state against external aggression, not to dabble in the political life of a people, employing violence to camouflage its ignorance of the complex forces that guarantee and sustain the human organism which we describe as a nation.

The sudden, unprecedented surge in the populations of those refugee camps mentioned earlier, boosted by the unspeakable dimension of genocide perpetrated in Rwanda, pulls us up sharply. It would rebuke as self-indulgence even the foregoing illustrations of failed nationhood except that they are intended as a warning, a cry for help in averting a similar catastrophe. Prophecy is not my line of business, but a hard question is being posed by events of unprecedented inhumanity, and not only in that benighted part of the world called Rwanda. As I speak, the routed Hutu army is regrouping, infiltrating, its notorious militia readying itself to complete its interrupted programme of 'ethnic cleansing'. The Rwandan Patriotic Front, Tutsi dominated, has demonstrated that its cadres are not above acts of vengeance. A vicious cycle is the inevitable legacy of generations yet unborn.

In such circumstances, is it really impossible to think the unthinkable? A scourge that terminates half a million souls in a matter of weeks is not an assault that is confined to the directly affected space; it sends ripples of dread through neighbouring and even distant spaces with a history of internecine conflicts. Suppress it how you will, the question that is precipitated today in a million thoughtful minds across the continent is: Could it happen here? And even those who have developed, over centuries or decades, a far more benign way of resolving such conflicts must surely, today, cast a nervous glance over other parts of the world and wonder how soon their resources might be taxed, and how severely, in order to rush aid to those sections of the world where yet another eruption makes urgent claims on their humanity.

*****

To those who have come to take certain social ideals for granted, or whose concerns are focused on the more readily intrusive ills of humanity, the clamour of voices for the enthronement of democracy as a condition of social existence throughout the world may sometimes sound a misplaced priority. Hunger is a more readily apprehended reality, and so is disease or exposure in any form to the unmediated ravages of nature. The sight of skeleton trails across the landscape of the continent, the swollen bellies of malnourished children, or images of entire villages in East Africa, devastated by the scourge of AIDS, appear far more pressing and demanding of global attention than some peripheral notion of governance called democracy. Indeed, we are accustomed even to the reduction of the problem to one of ethnicity; and, certainly, slaughter slabs such as Rwanda lend force to this circumscription of social complexities into easily digestible, even if unpalatable, capsules.

The result is that we are sometimes assailed by voices which have grown so insolently patronizing as to declare that Africans do not really care who governs them, or how, as long as they are guaranteed freedom from diseases, shelter and three square meals a day. I have answered this reductive proposition of the African political personality in other places. I do not propose to give one more second to such racial slurs, least of all when they are given voice from our own kith and kin, in possession of, or slurping from, the bloodied trough of power.

We propose instead that, much as those negative facets of existence contribute to social retardation in their own right, a common denominator of crisis in many areas may actually be found in the refusal of a section — be this understood in terms of class, ethnic grouping, profession or religion — to grant others the simple right of participation in the process of deciding a collective destiny. The military dictatorships of the African continent — parasitic, unproductive, totally devoid of social commitment or vision — are an expression of this exclusionist mentality of a handful. So are those immediately post-colonial monopolies that parade themselves as single-party states. To exclude the sentient plurality of any society from the right of decision in the structuring of their own lives is an attempt to anaesthetize, turn comatose, indeed idiotize society — which of course is a supreme irony, since the proven idiots of our post-colonial experience have been, indeed still are, largely to be found among the military dictators. I do not suggest that the level of intelligence of the military in general is any lower than that of the civil society; no, we have evidence to the contrary. I merely propose that it is the dregs who, against all natural laws, appear to rise to the top.

Under a dictatorship, a nation ceases to exist. All that remains is a fiefdom, a planet of slaves regimented by aliens from outer space. The appropriate cinematic equivalent would be those Grade B movies about alien body-snatchers. The only weapon of resistance that is left intact is a cultural memory. Make a careful study of peoples under a dictatorship and, invariably, you will observe that it marks a period of internal retreat into cultural identities. This process is entirely logical. The essence of nationhood has gone underground and taken refuge in that primary constituency of human association — the cultural bastion. The longer the dictatorship lasts, the more tenacious becomes the hold of that cultural nationalism, attracting to itself all the allegiance, the social relevance and visceral identification that once belonged to the larger nation. That is the painful lesson that the former Soviet empire has for the world. It is the continuing lesson of the Sudan, if the world would only remove the blinkers from its eyes and unplug its ears! Our acceptance of the centrality of the human entity in a rational social order assumes implicitly and comprehensively all elements that go into the making of that human entity, and these include intelligence and the capacity for choice. These attributes are violated from the moment of dictatorship, and the populace becomes, by definition, a collection of mutants. To regain their wholeness, they withdraw into the cultural sheath; it remoulds their psyche and restores their sense of worth.

A dictatorship does not, as we have seen, merely annul the process of choice and participation, which might take the form of an election. It annuls, effectively, the nation itself. If therefore there is an organization that claims to be a club of nations, what, may we ask, are dictators doing within such an organization? What nations exactly do they represent? A rubber, timber or mining concession granted to an entrepreneur for exploitation also involves the making and implementing of local laws for the survival and profits of the company. It may involve the construction of independent airfields, establishment of company police, public levies or other forms of taxation, capital-intensive and autonomous infrastructures, and so forth. The tin mining enclaves of Jos and the Plateau region of Northern Nigeria (until some years ago) or the contemporary hi-tech Brazilian jungle lords, violating virgin spaces and wreaking ecological devastation, are pertinent exemplars. But does any or all of this turn such spaces into nations? What makes the American businessman's private army in the Andean jungle depths any more alien to that environment than a power-crazed drug warlord, an indigene of the Nigerian or any other space, who happens to have succeeded in seizing control of a section of a nation's armed forces and cows the rest into submission through torture, threat of dismissals, imprisonments and secret executions? Is such a being a representative voice of a nation? It is time this question is invoked within the United Nations. All organisms evolve, and so must the United Nations.

*****

When we espouse the cause of democracy, then, our minds encompass more than the ritual of the polling booth and the change of baton at the end of an agreed number of years. Side by side with the eradication of the uniformed mutants who erupt from time to time on our national landscapes, we consider also a dispensation that enables all humanity to breathe freely, to associate freely, to think freely and to believe or not believe without a threat to its existence, and without discrimination in social rights. Implicit in that freedom of association is — difficult as it may be to accept — the right of collective dissociation. The Canadians, at the time of writing, have yet to resolve the question of their own single or dual nationhood, and no exponent of either choice has been advertised as some species of retarded development. The price that is being extracted from the people of Chechnya for their choice of association should be unacceptable to the twenty-first century mind. It is primitive, even prehistoric, to have unleashed such brutality on a section of a contested collectivity. Nothing justifies such a savagery of subjugation, carried out regardless of the lives of the innocent and non-partisan in this dispute. Just as primitive is the thirty-year long struggle of the Sudanese people to decide the conditions of their social existence and the primacy of their religious beliefs.

Therefore, despite the emphasis of most of the preceding on the internal realities that uphold or belie the claims of nationhood, and on the need for a redefinition of such a social organism, based on the conditions of its inmates and their relationship to power and leadership, I remain inclined to state a concluding conviction that the time has come for a structured pattern of regional conferences on the national question in numerous sectors of the globe — and most especially in those sectors whose propaganda machinery effectively denies the existence of such a question and whose state violence ruthlessly silences the restive inmates of what amounts to vast prison yards. While a number of such conferences may prove fruitless, it is quite possible that a few may actually succeed in heading off the debacle of Somalia, the horrors of Yugoslavia or the reversion to the hitherto unthinkable, primitive bestiality of Rwanda. It is time for self-defining nations to swallow their pride and redefine their assumptions — but only from the rigorous view of the well-being of millions who constitute their humanity, not from the perspective of games of power that constantly seek to aggrandize, to bloat the winners through dubious claims on resources, derived from mere spatiality, religious mandate, ethnic purity, or domination over potential killing fields.

The history of many nations is so flawed that it screams out constantly for redress. Let us therefore summarize the lessons of that history in all objectivity, keeping out the jingoism that attends the traditional line of thinking to which we are largely habituated. Neither the tenacity of state repression nor the longevity of an illusion is adequate to guarantee an eternity to a nationhood whose foundations are unsound, and whose super-structures, however seductive, are constantly stressed as much by the incubus of collective memory as by the dynamics of human development, in both its quantifiable and intangible aspects. I believe that the human mind can encompass this recognition in an original, revitalizing way, enabling us to map — literally — new directions that redress the history of societies and humanize the destiny of their peoples.

Positive Aspects of Community and the Dangers of Fragmentation

Amitai Etzioni

In addressing my fellow citizens of the developed countries of the West and the East, the thesis which I put before you is that the welfare state has exhausted its capacity to undertake an ever-widening array of social missions and to pay in full for those it has already shouldered, because it has exhausted its ability to raise taxes and is losing political legitimacy. Communitarians like myself do *not* oppose the welfare state in principle, and we join with many, including conservatives, who call for a responsible economy or 'social market'. Others refer to this as 'welfare capitalism', the notion that the state will continue to play a central role in providing social services, from health insurance to unemployment benefits. However, it seems that as new social needs are identified (such as a need for extensive nursing home services as the population ages), and as the costs of health care and other services continue to rise rather rapidly, it is not possible to expect the welfare state to expand much further. On the contrary, within many developed nations there are movements afoot to reduce the scope of benefits and services already provided by the state. Citizens of these countries might do well to ponder what I call the Seattle story.

Imagine that you are the top health commissioner of Seattle, Washington. You report to the mayor that new medical research strongly suggests that

This piece draws upon the author's *The Active Society* and *A Responsive Society* (Etzioni, 1968, 1991).

once a person has a heart attack he/she must be reached within four minutes
to prevent irreparable brain damage. This requires positioning an ambulance
every thirty-two blocks, which would cost the city millions of dollars each
year. The mayor responds, 'Sorry, I do not have that kind of money'.
Without the necessary funding, what would you do next?

You could do what Seattle did in reality; call upon individuals and
organizations to lend a helping hand. Since 1985, various private groups in
Seattle have trained 400,000 people, nearly half its citizens, in Cardio-
pulmonary Resuscitation (CPR). Now it is very likely that if you keel over in
Seattle, someone will be there within thirty seconds to assist you, without any
cost to the public. Low in costs, high in benefits, the act of helping one
another also serves to sustain the spirit of community, which in turn has
many indirect beneficial effects — although none as dramatic as turning a
breathless, blue body into a fully functional fellow human being.

In Seattle, people feel better about themselves and others than in more
atomistic urban environments, such as New York City. They feel less
isolated (in part, because they meet in various CPR training and refresher
classes). Moreover, their communitarian spirit spills into other areas; for
instance, Seattle has one of the highest levels of voluntary recycling in the
United States. Last but not least, these are the ways in which people develop
their civic 'muscles', because participation in such social activities is a seed
bed for cultivating democratic practices.

In the United States, much public good is done through such means,
ranging from several thousand neighbourhood crime-watch schemes and
patrols, to ethnic groups that help immigrants of their 'own kind', to
associations that organize block parties and other social events. I suggest
that, in the future, developed societies will need to do more societal business
following this model, because there is little evidence that the economies of
developed nations will shift into higher gear in the foreseeable future,
enabling them to generate large amounts of additional public resources from
increased tax revenues. Hence the debate that is often waged between liberals
(who favour reliance on the state) and *laissez faire* conservatives (who believe
in the market) is beside the point. Both overlook the communitarian sector
that we must increasingly rely upon in the coming years.

COMMUNITARIANS AND THEIR CRITICS: THE GROOVED DEBATE

The intellectual debate about the foundations of communitarian thinking is
stuck like a gramophone needle in the groove of an overplayed record. For
decades now, communitarians have been pointing out to libertarians[1] that

---

1. The terms 'liberals', 'classical liberals', 'contemporary liberals', and 'libertarians' have all
    been used to characterize the critics of communitarians. These labels are confusing; for

individuals are not free-standing agents but members of communities. While people survive without communities, the thinner their community bonds, the more alienated and unreasoning they tend to be. Moreover, because for communities to flourish they require that their members not be completely self-oriented, the common good has a normative standing in the same sense that life and health do: they all are essential for our physical and spiritual well-being.

Libertarians in turn have either simply ignored these arguments, with talk of the choices which individual consumers, voters, or others make on their own; or else they have come to depict communities as social contracts, something which free-standing individuals construct because it suits their individual purposes. Libertarians seem afraid that the recognition of the common good as a value that is equal with personal freedom will endanger the standing of that liberty.[2] Nevertheless, as Talcott Parsons pointed out in his discussion of pattern variables, while value systems can maximize one dimension or theme, societies — which must attend to a variety of conflicting requirements — are inevitably organized by several principles. They must concern themselves both with order *and* with the freedom needed to search for new adaptive patterns of social order. They must take into account both the justice of allocations *and* productivity, and so on. In short, ideologies and ideologues can afford to be, in fact even benefit from being, one-dimensional. Students of society should know better.

The Responsive Community

In the following pages, I shall build upon the composite concept of the 'responsive community', based upon the assumption that there is an irreducible, partially productive, and tensed relation between the centripetal forces of community and the centrifugal forces of autonomy. The discussion deals with the sociological equivalent of ordered liberties, rather than suggesting a libertarian free-for-all. First, the term 'responsive community' suggests that individuals are members of one another, that people are

---

instance, many readers do not realize that the labels are not confined to or even necessarily inclusive of those who are called liberals in typical daily parlance. Most importantly, because the defining element of the position is the championing of the individual, 'libertarian' seems both the less obfuscating term and the one that is substantively most appropriate.

2. Among the recent books that cover this debate extensively and well are Selznick (1992) and Bell (1993); see also Phillips (1993). In addition, the topic dominates collections of essays by Daly (1994) and Benhabib (1992). These collections suffer from a tendency common in public debates — the tendency to polarize. Social theorists, philosophers, and writers are readily labelled as liberal or conservative; there simply seems no room, or no inclination, to recognize other, more nuanced positions.

ontologically embedded in a social existence. Second, because all that is valued by members of a society, liberty included, is dependent on sustaining the social realm, a measure of commitment to the commons has a moral standing. Third, I argue that it is not possible, desirable or morally justifiable fully to absorb members' identities, energies and commitments into the social realm. Providing for individual liberties limits the costs of maintaining social order, allows members of society to express the idiosyncratic aspects of their selves, and enables the development of new social patterns that are more adaptive to the ever-changing external environment and internal balances of communities than traditional patterns might be.

The term 'responsive communitarian' serves not only to distinguish this position from the libertarian one (which builds on free individuals as its conceptual and normative cornerstone), but also to distinguish it from those communitarians who build on the empirical and ethical significance of the community without attention (or with insufficient attention) to individual rights. One example of the latter is Alisdair MacIntyre, who dismisses rights as 'fictions' (1984: 70). The term 'responsive' implies that the society is not merely setting and fostering norms for its members, but is also responding to the expressions of their values, viewpoints and communications in refashioning its culture and structure.[3]

Taken together, these three assumptions suggest that the starting point — the primary concept of responsive communitarian thinking and one that should be able to advance the communitarian–libertarian debate beyond its current groove — is the concept of a permanently tensed relation between individuals and the society of which they are members. Centrifugal forces will tend to lead individuals to break out, dangerously reducing the social realm, in their quest for ever more attention to their particular, individual, sub-group agendas. Centripetal forces will tend increasingly to collectivize members' energies in the service of shared goals, and to curb their degrees of freedom. A society functions best when both forces are well balanced. Nevertheless there tends to be a continual tug of war between them, and the society and its members are constantly pulled in one direction or another. It is the role of social observers and commentators, of intellectuals, to establish in which direction society is leaning and then to throw their weight on the other side of history. Thus in contemporary China, Albania and the former Soviet Union, the intellectual case is for better anchoring of individual liberties; in the contemporary United States, the case is for more commitment to the common good. In other words, while responsive communitarians within any one given societal context or historical period may argue for more community (as in the present-day United States) or for more individual rights (as in present-day China), they

---

3. For a further discussion of this issue, see Etzioni (1968, 1991).

actually seek to maintain the elementary balance that is at the foundation of all good societies.

We are now ready to face several important subsidiary issues that arise once this basic position is accepted. Those who attack community argue that the term community is ill-defined; that communitarians are nostalgic and overlook the darker sides of community, or even that they wish to retrieve those less appealing features; that communities, as communitarians define them, never existed or are not sustainable in a modern society; that communities are inherently majoritarian; and that communities oppress their members. Thorough discussion of each of these issues could fill a volume; but here I will respond at least briefly to some of the challenges put forward by critics.

Community Ill-Defined?

In 1991, Robert Booth Fowler devoted a whole book to the confusion that may be associated with the concept from which communitarians derive their name. As he sees it, the term 'community' is not well-defined; indeed it harbours a large variety of meanings (he analyses six). 'The picture is thus confused and complicated. The meaning of community is elusive, a word without an essence or a text without meaning' (Booth Fowler, 1991: 3). Some who are less sympathetic to the idea of community make the same point. Stephen Holmes (1993: 177) asks, 'But what is community? What does it look like?'. He finds the communitarian answers to these questions inadequate. Jack Crittenden (1992: 136) claims that communitarians avoid difficult questions concerning community by 'remaining vague about the nature of community'.

Actually the term community can be defined with reasonable precision. It refers to a group of people who share affective bonds and a culture, and it is defined by two characteristics. First, communities cannot exist without a web of affect-laden relations among a group of individuals (rather than merely one-on-one, or a chain of individual, interactions). These relations often criss-cross and reinforce one another. Second, being a community entails having a measure of commitment to a set of shared values, norms and meanings.

The Darker Side of Community?

Some of those who object to the call for a return to community accuse communitarians of being nostalgic and/or conservative. The first accusation is that communitarians have a rosy view of the past. Those who long for community, this argument goes, conveniently ignore the darker side of traditional communities. 'In the new communitarian appeal to tradition, communities of "mutual aid and memory", and the Founders, there is a problematic inattention to the less attractive, unjust features of tradition', writes Linda McClain (1994: 1029). The charge of conservatism accuses communitarians not of overlooking the less attractive features of traditional communities, but of wilfully longing to retrieve those features. According to Michael Taves, the communitarian vision of community rejects modernism, concerning itself mostly with 'reclaiming a reliance on traditional values and all that entails with regard to the family, sexual relations, religion and the rejection of secularism' (Anderson et al., 1988: 8). Katha Pollitt, after discussing the communities that communitarians wish to retrieve, concludes that communitarianism is a conservative movement masquerading as something else: 'Communitarianism [is] Reaganism with a human face' (Pollitt, 1994: 118). As Booth Fowler puts it, these critics 'see talk of community as interfering with the necessary breaking down of dominant forces and cultures'. Community is viewed as inherently traditional and conservative.

These criticisms are sound but misdirected. Early communitarians might be charged with being in effect social conservatives, if not authoritarians. However, the new school of responsive communitarians, including scholars such as Charles Taylor, Philip Selznick, Robert Bellah, Thomas Spragens, William Galston and (less directly involved) Michael Walzer and Michael Sandel, fully realize and often stress that they do not seek to return to the kind of traditional community which has been associated with discriminatory practices against minorities and women, an authoritarian power structure and rigid stratification. Responsive communitarians seek to build communities on open participation, dialogue and truly shared values. (To be fair to McClain, she recognizes this feature of the responsive communitarians, writing that some of the latter do 'recognize the need for careful evaluation of what was good and bad about [any specific] tradition and the possibility of severing certain ... features from other [features]', McClain, 1994: 1030.) As political scientist R. Bruce Douglass writes (1994: 55): 'Unlike conservatives, communitarians are aware that the days when the issues we face as a society could be settled on the basis of the beliefs of a privileged segment of the population have long since passed'.

Never Existed, Incompatible with Modernity?

Another group of community critics, meanwhile, argues that it is impossible to lose or to reform what we have never had, and that community is not in any case sustainable under conditions of modern life. 'Communitarians have been mistaken in their claims about the prominence of community in times gone by', writes Derek Phillips (1993: 175). He adds, lest he be misunderstood, 'If the sort of community depicted by communitarian thinkers did not exist in the past, then it obviously cannot be said to have given way to the forces of modernization' (ibid: 149). In what is known in legal circles as the kitchen sink defence ('If the first line does not work, I will try this one'),

Phillips adds to his attack: 'Even if community was once widespread, that does not mean that it is a viable option today' (ibid: 8).

The high rate of geographic mobility of modern society is the obstacle most often cited by those who question the current feasibility of community. Ken Anderson maintains: '[W]hen push comes to shove, most people are not so enamored of community as they are of mobility' (Anderson et al., 1988: 22). David Seeley (1991) believes that the communitarian vision 'smacks of a Norman Rockwell America that no longer exists and, perhaps, never did. More Americans live in faceless apartment buildings, condos and housing tracts than in towns where rustic but decent folk gather regularly to speak their minds'.

The fact is that communities are not relics of a pre-modern era. While it is true that modern economies entail a highly mobile society, people have learned to develop community bonds relatively quickly. Moreover, many communities are not residential and hence provide a measure of adaption to mobility. A member of the gay community who moves to a different city is likely to know personally some individuals in the new city, or at least to meet some. He will be able to find the core institutions in which his community congregates, and will be familiar with the basic elements of its culture, norms and meanings. He will be relatively quickly integrated into the local embodiment of his community. The same holds for a Jew (who is likely to be a member of a sub-community of orthodox, conservative, or secular Jews), for Korean-Americans, for Cuban-Americans and for many others.

Majoritarian?

Communities are said to be 'majoritarian' when — to the extent that the community has shared public policies — the majority determines the course to be followed. This argument comes in numerous forms. Nadine Strossen, a professor of law at New York University and the president of the ACLU, sees communities as threats to minorities, claiming that communitarians are 'majoritarians' who are willing to tinker with the rights of minorities (quoted in Scheinin, 1993: 3). Tibor Machan sees in the communitarian conception of community a threat to the individual. 'Communitarians wish to place community and individual on a collision course, saying there is some kind of balance that is needed between the rights of individuals and the rights of community ... But if we consider that community means simply a lot of people other than oneself, this simply makes for majority rule' (Machan, 1991). According to Charles Derber, the consensual values that are a crucial aspect of community are also potentially majoritarian, as these values are simply 'the voice of one part of the community — usually the majority or an elite minority — against the others' (Derber, 1993: 29).

Michael Sandel (1990: 155) offers a response to these charges. 'The answer to that majoritarian threat is to try to appeal to a richer conception of democracy than just adding up votes'. American society has both constitutional and moral safeguards against majoritarianism that communitarians very much respect. These safeguards basically work through differentiation, by defining some areas in which the majority does not and ought not have a say, and others in which it does and should. The United States is not simply a vote-counting majoritarian democracy, but a constitutional democracy. That is, some choices, defined by the US Constitution, are beyond the realm of the majority. Clearest among these is the Bill of Rights, which singles out matters that are exempt from majority rule and from typical democratic rule-making. The First Amendment, which protects the right of free speech, is a prime example of an area in which minority and individual rights take precedence. Similarly, the majority may not deny any opposition group the right to vote; even Communists were not banned in the days when they were most hated and feared.

The US Constitution and American legal traditions and institutions indicate clearly, however, that other matters are subject to majority rule. Thus majorities decide how much tax Americans must pay, which side of the road to drive on, and at what age young adults can vote. There is neither moral nor legal support for the notion — indeed it is inconceivable to believe — that individuals could decide for themselves whether or not to pay social security taxes, which side of the road to drive on, and so forth.

Culturally Oppressive?

In a criticism that combines the allegation of conservatism with that of traditionalism, critics have argued that communities — even when they do not use coercion — may strongly pressure their members to abide by a culture that the members do not truly share. According to Will Kymlicka (1993), this oppression can entail prescription by the community of forms of subordination, roles that limit people's individual potential and threaten their psychological well-being. Similarly, Michael D'Antonio (1992: 50) claims that communitarians ignore the 'stifling pressure to conform that is usually present in ... close-knit settlements', while Amy Gutmann (1985: 319) proclaims that communitarians 'want us to live in Salem'.

From a contemporary responsive communitarian view, the combination of a concern with traditionalism and cultural oppression is not accidental. Traditional communities were often both (hence the well-known line that 'the air of the cities frees', which is what the farmers of traditional villages must have felt when they first moved into cities at the onset of industrialization). There is neither need nor reason to deny that some totalitarian communities exist in our time. Nevertheless contemporary communities are typically not exclusive, even when they are territorial, which they often are not. A person may be a member of a suburban community and a work community, a residential community and an ethnic one, and so on. The result is that each community has less of a stronghold on the person than might otherwise be the case. The relative ease of mobility in our times indicates that people choose which community to join — and in which to continue. In short, the problem of most contemporary communities in pluralistic, democratic societies is not that they are overpowering, but rather that they are anaemic.

There is, however, a much more challenging issue hidden here. It is sometimes touched upon in conjunction with the charge that communities promote conformism, even of the weaker kind. The question concerns the normative/moral standing of whatever values the community promotes, however gently it may promote them. (Note that all communities share some values by definition and tend to promote these shared values to sustain themselves.) Should members abide by them? On what grounds? Under all conditions? Some liberal communitarians solve this problem by arguing that communities should not have one shared characterization of the common good, but rather they should maintain a plurality of conceptions. In this way, each person could choose the values to which he or she would subscribe. Nevertheless this assumption flies in the face of sociological fact; a shared set of core values is necessary for a community to be able to build consensus on specific norms and policies — in short, to function as a community.

It follows that both members of communities and outside observers need to evaluate the moral standing of the values of any given community, rather than merely endorse them because they are shared, because they exist, or on some other such utilitarian ground. One criterion for evaluation might be whether or not the vision of a community, and of the future, which these values underwrite fully meets the authentic needs of all the members of the community. Reference is made to all the needs of the members because responsive cultures are encompassing, attending not merely to all the members but to all their basic human needs.

Finally, the term 'authentic' calls attention to the possibility that extensive propaganda may make members of a community support values to which they are not truly committed. The term further suggests that a community need not be responsive to surface demands, but only to true ones. The ways to distinguish authentic from false preferences cannot be discussed here in any detail; but suffice it to say that when people are falsely committed, this is evidenced by their behaviour, above all by strong tendencies to return to their authentic values whenever propaganda and state controls slacken, and by extensive attempts by members of the community to circumvent the behaviour sanctioned by false values.[4]

Many may find this position hard to accept on the grounds that it raises empirical difficulties (for instance, how to assess authenticity) and normative ones (especially for those who seek to challenge the shared values of the

---

4. For further discussion on this, see Bell, 1993; Etzioni, 1968; Goodin, 1991.

community in the name of one sub-group or another). But the issue cannot be avoided. Communitarians must be able not only to account for the fact that communities share values but also to respond to the following question: Is it morally appropriate to be guided by the particular values that are shared? On what criteria is one to draw in evaluating values?

In short, while responsive communitarians seem to have reasonably sound replies to many of the criticisms raised against them recently, the standing of community values is the one that needs the most attention and so far has received the least. More broadly speaking, communitarians and their critics must break out of the grooved debate about the socially embedded self and turn to the numerous issues raised subsequently. None of these challenges the basic communitarian thesis — that communities, properly constructed, are of great value.

Unitary Structures and Intolerance: The Threat to Democracy

An extremely important issue for those concerned with promoting responsive communities is posed daily by events like the breakup of the former Soviet Union and Yugoslavia, threats of secession in Canada and India, and many other manifestations of the centrifugal forces which can be unleashed by the search for ever-more-exclusive, ever-more-homogeneous forms of community. Those who foster community building must realize that the stronger communities become, the greater the danger that they will turn hostile to outside groups. This is not a sufficient reason to oppose communities *per se*, but it is a strong reason for ensuring that they remain constructive in their dealings with one another. It is, therefore, worth our while to examine the tendency for existing communities to break into ever smaller ones in the name of self-determination, and to explore the difficulties involved in granting unquestioning moral support to this important principle. The relation between this issue and others discussed above cannot be ignored.

Most nations in the world contain numerous ethnic groups, within which further cultural differences exist. Moreover, additional ethnic 'selves' can be generated quite readily, drawing on fault lines now barely noticeable. Centrifugal forces are always present, as Dov Ronen (1979: 8) notes:

> Because the new 'us' often becomes just another framework that appears to limit the freedom of the individual, of the real 'self,' the perception of a new 'them' is prompted, and hence the formation of a new 'us,' for the further pursuit of the aspired-to 'freedom' and 'good life.' And so a new quest for self-determination evolves, with another new 'us'; and then another, possibly *ad infinitum*.

The phrase *reductio ad absurdum* never had a more meaningful application.

There are, of course, situations in which national frontiers bind people into patently unworkable or tyrannical political systems, but the breakup of large entities does not necessarily provide for movement in the direction of democratization. In fact, excessive insistence on self-determination can weaken countries seeking to establish democratic government and threaten democracy in countries which have already attained it. There are at least two reasons for this, one structural and one socio-psychological. The first is related to the merits of pluralism; the second, to the importance of tolerance. It is necessary to explore the two arguments separately, although there is a strong connection between them.

The Merits of Pluralism

To ensure that the variety of needs within the population will find effective political expression, democracies require that the government in place not 'homogenize' the population in some artificial manner. It is the plurality of social, cultural and economic loyalties and power centres within society that makes it possible, at any point in time, for a new group or sub-culture to break into the political scene, find allies, build coalitions and achieve some of its goals. As well as keeping the government and its closest allies in the population in check, pluralism allows many groups to counterbalance each other's interests. In contrast, when historical processes or deliberate government policies weaken all groups except those supporting the régime, as in the case of the Nazis in Germany following the First World War, the foundations of democracy are undermined. In short, *social pluralism is a major factor that supports democratic government.*

While there are several bases upon which pluralism can be sustained, the best are those that cut across other existing lines of division, dampening the power of any single group and creating a large number of possible combinations on which the bases of political power can be built. Thus a society rigidly divided into two or three economic classes (say, landed gentry, bourgeoisie and working class) may have a structure that is somewhat more conducive to democratic government than a society with only one class. Nevertheless the potential for democracy is much enhanced when there are other groups that draw on members from various classes, so that loyalty to these groups cuts across class lines.

Historically, ethnic groups have 'cut across' socio-economic levels within the United States, thus dampening both class and ethnic divisions. Thus American Jews may be largely middle class, but there are many in the middle class who are not Jewish, and there are Jews in other classes. WASPs (White Anglo-Saxon Protestants) may be 'over represented' in the upper classes but are also found in large numbers in all other classes, and so on. The fact that both class and ethnic loyalties cut across regional, geographical and other lines, helps to further cement the foundations of pluralism and hence democracy.

In contrast, the formation of break-away states based on ethnic groups creates communities which are sociologically more monolithic than the states from which they have separated. Thus, to take the Canadian example, a secessionist Quebec would obviously be much more 'French' than the existing composite, and the remaining Canadian nation more 'English'. Polarization can be heightened by the greater intolerance which break-away states tend to have for minority ethnic groups composed of people who were in the majority or in power in the original state. This has certainly been the case for Russians remaining in many of the newly formed republics of the former Soviet Union. In short, ethnically based break-away states tend to be characterized by more ethnic homogeneity and less pluralism, and this is one factor which may undermine the deeper sociological foundations of democracy.

The Role of Tolerance

Democracy requires tolerance (whether of people of a different background, sub-culture, religion or language, or simply of people with different political views) because this is the psychological basis for playing by the rules, for being willing to accept the outcome of elections — even if they favour a party or coalition of groups to which one is strongly opposed — and for accepting compromise. The creation of broad-based communities requires the same basic psychological predisposition: the capacity to respect people of different backgrounds and traditions; the ability to work out differences with people whose religions, histories and habits one does not share. When this predisposition is absent, the risk of fragmentation is high.

Since the ultimate purpose of self-determination is not self-determination *per se* but a government responsive to those governed, it can be argued that before taking up the banner of self-determination, every effort should be made to reform existing governmental structures, to render them more responsive and tolerant. Only if such efforts fail does there seem to be justification for an ethnic group to break away. When this last-resort course is embraced, the newly formed communities should take special pains to develop tolerance, lest they become even less democratic and less responsive to major groups of their citizens than the nations from which they have separated.

While there is nothing wrong with basing community on ethnicity (and a great many communities are so based), there is often much to be lost when ethnicity becomes the defining characteristic of a state. Much may also be lost if an attempt is made to erase ethnic identity in pluri-cultural societies (what Arthur Schlesinger, Jr. [1991] calls applying the great 'solvent' of total assimilation). Ethnic groups should continue not only to exist, but to thrive and enrich the cultural mosaic, within the context of democratic pluralism.

REFERENCES

Anderson, Ken, Paul Piccone, Fred Siegel and Michael Taves (1988) 'Roundtable on Communitarianism', *Telos* 76: 2–32.

Bell, Daniel (1993) *Communitarianism and Its Critics*. Oxford: Clarendon Press.

Benhabib, Seyla (1992) *Situating the Self: Gender, Community and Postmodernism in Contemporary Ethics*. New York: Routledge.

Booth Fowler, Robert (1991) *The Dance With Community: The Contemporary Debate in American Political Thought*. Lawrence, KA: University of Kansas Press.

Crittenden, Jack (1992) *Beyond Individualism: Reconstituting the Liberal Self*. New York: Oxford University Press.

Daly, Markate (1994) *Communitarianism: A New Public Ethics*. Belmont, CA: Wadsworth Publishing Co.

D'Antonio, Michael (1992) 'Tough Medicine for a Sick America', *Los Angeles Times Magazine*, 22 March, pp. 32–50.

Derber, Charles (1993) 'Coming Glued: Communitarianism to the Rescue', *Tikkun* 8(4): 27–30.

Douglass, Bruce R. (1994) 'The Renewal of Democracy and the Communitarian Prospect', *Responsive Community* 4(3): 55–62.

Etzioni, Amitai (1968) *The Active Society: A Theory of Social and Political Processes*. London: Free Press.

Etzioni, Amitai (1991) *A Responsive Society: Collected Essays on Guiding Deliberate Social Change*. Oxford: Jossey-Bass Publishers.

Goodin, Robert E. (1991) 'Permissible Paternalism: In Defense of the Nanny State', *Responsive Community* 1(3): 42–51.

Gutman, Amy (1985) 'Communitarian Critics of Liberalism', *Philosophy and Public Affairs* 14(3): 308–22.

Holmes, Stephen (1993) *The Anatomy of Antiliberalism*. Cambridge, MA: Harvard University Press.

Kymlicka, Will (1993) 'Some Question about Justice and Community', in Daniel Bell *Communitarianism and Its Critics*, Appendix. Oxford: Clarendon Press.

Machan, Tibor (1991) 'The Communitarian Manifesto', *Orange County Register* (12 May).

MacIntyre, Alasdair C. (1984) *After Virtue: A Study in Moral Theory*. Notre Dame, IN: University of Notre Dame Press.

McClain, Linda (1994) 'Rights and Irresponsibility', *Duke Law Review* 32: 989–1088.

Pollitt, Katha (1994) 'Subject to Debate', *The Nation* 259(4): 118 (25 July).

Phillips, Derek L. (1993) *Looking Backward: A Critical Appraisal of Communitarian Thought*. Princeton, NJ: Princeton University Press.

Ronen, Dov (1979) *The Quest for Self-Determination*. New Haven: Yale University Press.

Sandel, Michael (1990) in Bill Moyers *A World of Ideas: Volume 2*, pp. 149–57. New York: Doubleday.

Schlesinger, Jr., Arthur (1991) *The Disuniting of America: Reflections on a Multicultural Society*. New York: Whittle Communications.

Seeley, David (1991) 'Mushy Thinking on Individual Rights', Letter to the Editor, *Wall Street Journal* CCXVIII(88): A15 (1 November).

Selznick, Philip (1992) *The Moral Commonwealth: Social Theory and the Promise of Community*. Los Angeles, CA: University of California Press.

Scheinin, Richard (1993) 'New Philosophy Urges a Social Commitment: Critics Say Communitarians Sacrifice Individual Rights', *The Press-Enterprise* (30 January) Section G, p. 3.

The Perils of Utopia: The Russian Intelligentsia under Communism and Perestroika

Tatyana Tolstaya

If Gorbachev's perestroika turned out to be much more painful for the Russian intelligentsia than might have been expected, there were profound, long-standing reasons for this. But before examining these, we must first ask: what is the Russian intelligentsia? Who are these people? A precise definition remains elusive: like others before me, I have often tried to define the intelligentsia, each time unsuccessfully. These sorts of discussions and debates were once popular in Russian literary and culturological publications. I was among those invited to participate in them and, in an attempt to be precise in expressing my understanding and views, I usually ended up completely muddled. However, I did eventually come to understand something: why it is so difficult, if not downright impossible, to say exactly who comprises the Russian intelligentsia.

The problem lies in the very structure, the chemical composition, if you will, of the Russian cultural fabric, which is so different from its Western counterparts. The very concept of 'definition' is redolent of the kind of logic, the scientific spirit and rationality that prevents one from seeing the emotional chaos which engendered the philosophical spirit that transforms some Russian people into what we call the intelligentsia. When we attempt to impose a stable definition on this state of mind, the fabric under investigation disintegrates: the instrument destroys the object of its study.

Translated from Russian by Jamey Gambrell.

In applied Soviet Marxism[1] — that is, the Leninist recipe baked in the Stalinist oven — the intelligentsia was understood as a layer between the thick crusts of the 'genuine' classes: the workers and peasants. This culinary metaphor is not a slip of the tongue, but an image from a system of imagistic equations, one of the keys to a metaphoric code. What about the expression 'the cream of society'? What about Lenin's fantasy that 'every *cook* should know how to run the government'? It's as if a cannibal were preparing for the evening meal.

The battering of our population began with the intelligentsia, with those people to whom this label was applied, among them people who didn't count themselves part of it. Everything that was considered the best in society and culture was declared the worst and systematically destroyed. The word 'intelligent' became a curse, and the qualities of the 'intelligentsia' were eliminated everywhere. This was not just a matter of education, the acquisition of knowledge: in fact, under the Soviet regime, the purely formal acquisition of professional knowledge in all areas was welcomed, encouraged and propagandized. Literacy became almost universal, schools and institutes opened everywhere; nobody could deny the high level of Soviet science, especially fundamental science.

But none of this has any relationship to the intelligentsia and its qualities. Education and the intellect were not the sole ingredients of the Russian intelligentsia. It was distinguished by social concern (although in extreme cases it may have been a superficial concern, what in Russia was ironically referred to as 'social grief'), a feeling of responsibility for directing social processes; a sense of conscience toward one's fellows given the cruelty of the law and the indifference of the authorities; liberalism; belief in democracy and progress, in the 'moral law in the breast', as Kant put it; and the desire for knowledge — for others as well as one's self. The intelligentsia was characterized by ideas of universal human brotherhood, selflessness to the point of asceticism, active compassion to the disenfranchised, a Christian sense of service combined with an almost obligatory atheism. Of course, in order for these qualities to arise and develop, education is important; but enlightenment is even more crucial. While these two concepts overlap to a certain extent, they diverge even more strikingly. It is possible — as we see every day — to be a well-educated person, and at the same time a thorough scoundrel. It is possible to be illiterate, not even knowing the alphabet, but to have an enlightened soul, a humanistic system of values, what we call 'the heart's mind'.

So this multitude of people, from all classes and estates, including educated people with no heart, enlightened people with no education, and

---

1. I specify 'applied Soviet' because I don't want blithely to dismiss Western Marxism, against which I have a rather strong prejudice for obvious reasons, but which I neither know nor care to know.

those who happily combine the two — all of these people together comprise the many-headed community from which the intelligentsia is born, but by themselves do not equal the intelligentsia. The intelligentsia is out there in that crowd, but it is not exactly the same thing.

I will try to explain my thoughts with examples. There is a thin line between a person's professional, social status and his or her human, spiritual — humanitarian, if you will — essence. The idea of the intelligentsia glimmers somewhere along the axis of this line, with oscillations in both directions. The teacher, doctor and priest are the traditional triad of the Russian intelligentsia. The teacher and priest are clear, but what is the doctor doing here? What do these three have in common? Apparently it is the fact that all of them are called upon to understand, help and improve people, each in his own way. For that reason, in old Russia, defence attorneys were always considered members of the intelligentsia, while prosecutors were not. For the same reason, the artist and the ballerina are part of the intelligentsia, since they elevate the soul through art. In the West, neither the lawyer nor the ballerina has any claim on being considered intellectual — they do just fine without that. In Russia, emotion is always placed before the intellect. In Russia the police force is considered — always and unconditionally — as a profession antithetical in every way to the intelligentsia; it's seen as a dull-witted executive power, whose role is to repress and punish. Judges can be forgiven, but the police cannot, although their functions are obviously related. Russians expect everything from an *intelligent*, whatever his or her profession: they expect a certain wholeness and a multi-faceted development of knowledge and understanding that one would not normally associate with a profession. This can reach extremes of naïvety. I remember one incident that struck and touched me. On arriving in the United States in 1988, in the town of Richmond, Virginia, I was probably the first Russian who was neither an emigré nor an official, the first normal Russian person, who had been there for a very long time. Nostalgic emigrés came from far-flung towns for my talk. After the talk a couple came up to me. They had emigrated from Kharkov some time before. The middle-aged woman immediately started complaining to me about Americans: how superficial they are, how rationalistic, calculating, unspiritual — the usual Russian complaints about foreigners. 'You just can't imagine how little spirituality there is here!' the woman cried, warming to her theme. 'I went to the doctor because I have a bad liver. I'm lying there and the doctor is palpating my stomach. I ask him: "What do you think about Toulouse-Lautrec?" He says "What?" I ask him again: "What do you think about Toulouse-Lautrec? Do you like his painting?" And can you imagine what he said to me? "What does Toulouse-Lautrec have to do with anything? I'm treating your liver, just lie there quietly and don't talk". And this is a doctor! A DOCTOR! What sort of country is this!'

I couldn't think what to say to her. But I understood what was going on in this woman. The most important thing to her wasn't her liver — to hell with the liver; something is bound to hurt when you're middle-aged; it's only the body, you can't heal it. It was her soul that was seeking support — conversation, beauty, understanding — and in her naïvety, she went to the doctor in search of it, as one might have in Russia. Her soul was hurting, not her liver. In short, everything connected to teaching, lucidity, enlightenment, sympathy, cleansing, knowledge — any kind of spiritual development at all — has some relationship to the idea of the intelligentsia. It's almost religion, but without God.

*****

Returning to the culinary metaphor that I used earlier, the intelligentsia isn't a layer: it's the rum that soaks the cake of society. That is why, when the intelligentsia came under attack after the Revolution, it wasn't the educated class that was destroyed (indeed, the new authorities began building an educated class themselves); it was the spirit, meaning and essence of the intelligentsia that was wiped out — and they made a rather good job of it. The paradox of the situation was that the Russian intelligentsia wanted the revolution, facilitated it, participated in it, and heroicized and romanticized revolutionaries of all sorts, seeing in them the bearers of the spirit of change, progress, freedom. Lenin himself, in all outward respects a member of the intelligentsia (the grandson of a doctor, the son of a school inspector, a lawyer by education, a man who knew foreign languages, philosophy, history, and a writer who produced over fifty volumes in the course of his fifty-four years), was not an *intelligent*. But who could deny his intellect?

Of all the frightening statements that could be extracted from his writings, especially the late writings, one phrase in particular has always remained with me. In a telegram to one of his cannibal-colleagues he writes, worried: 'We're not shooting enough professors'. My brain stops short and freezes at this phrase. Not shooting enough — so we are shooting, constantly. But — not enough. We should shoot more. 'Professors' is a general word. Which professors? Who are these people? What are their political views? Are they plotting against the new regime? What can they do, how can they harm it? After all, they aren't soldiers, they don't have arms, they aren't manufacturing bombs, as a young student might, a romantic-terrorist. A professor in Russia at that time was a person of a certain age, a solid, family man, a scholar, a person of merits. So why shoot him? What can a professor do? He can be indignant at the horrors being committed, at the cruelty, the pogroms, he can privately register his moral protest against the violence — and so he must be swept away. That's it. All of them must be destroyed as bearers of the spirit of the intelligentsia.

It's interesting that when Lenin had had his fill, in about 1922 if I'm not mistaken, he sent 400 stubborn, unredeemable professors, philosophers and writers abroad, in order to rid the country of their unhealthy spirit.

There is no need to retell the horrors of the Soviet regime. Just one reminder: before the revolution, when the death penalty was revived after a hundred-year hiatus and several people were executed, the intelligentsia of the entire country was up in arms and called for revolution. After the revolution, when the famous political trials began, especially in the 1930s, the 'intelligentsia', or what was then so called, demanded that vigilance be heightened and that the 'bloody dogs' be executed. And they were executed, and so were members of the intelligentsia, at least everyone who could be classified as such.

Despite the fact that the intelligentsia was a spirit and not a profession — or perhaps precisely because of this — it proved impossible to uproot it completely. The word, so to speak, once again turned out to be harder than bronze and more enduring than the pyramids. The traditional professions of the intelligentsia were occupied by people of the new, servile-totalitarian spirit, so the intelligentsia retreated, dispersed into other, unnoticed, simple professions. This process reached its apogee in the 1970s, when in large and small cities all the elevator and boiler operators, janitors, night watchmen, and so forth, were unrecognized, homegrown poets, philosophers, artists and scholars, and it was impossible to get one of these low-paying and seemingly unprestigious jobs. Thought, art and the word retreated quite literally underground. Like radioactive waters, the intelligentsia seeped down through the earth and permeated the regime with its spirit, thereby eroding it.

Of course, this was hardly the same intelligentsia as before the war. It's not surprising: the regime continued to sniff out and uproot all signs of the old intelligentsia's spirit. At the end of the 1970s almost all the dissidents had been put in prison or sent out of the country. The regime grew sweaty and tired. It had no strength left to fight with this remaining, pale, bohemian fluorescence of the intelligentsia. But in its fruitless struggle, the regime undoubtedly did damage to itself; and if I may be allowed to play with culinary metaphors once more, I would put it this way: supposing that the intelligentsia is a layer of jam or cream, the authorities tried to scrape away with a spoon. They dug deeper and deeper into the cake, but the intelligentsia turned out to be rum, and there was no way to get it out of the cake. The authorities themselves became saturated with the alcoholic spirit. Their heads began to spin and their reason to stagger.

I remember another scene from the end of the 1970s. In Russian cities everyone's main tormentor, their avenging angel, was the plumber. In the city you always lived in a state of existential alarm: if something happened to the water in your apartment, you were immediately reduced to a humiliated, helpless, pathetic state. Only the plumber could save you, but he was elusive, drunk, capricious, he'd come only if he wanted to, or if the other people important to the functioning of your building — the elevator operator, the dispatcher, the janitor — could somehow persuade him. You had to make friends with him, be attentive to details, give him useful things, such as empty bottles that he could redeem to make a little money. One of my former colleagues, an editor at a publishing company, discovered problems with her plumbing. She collected a whole bag of empty bottles and went down into the depths of the building to the little room of the elevator operators and concierge in order to win them over and ask them to facilitate the appearance of the plumber. 'And what do I see', she told me the next day, 'but a pretty young elevator operator who's lounging on the sofa with her hair down. In front of her there's an easel, and an artist is painting her portrait in oils. On the table there's a typewriter and a stack of Ancient Greek dictionaries ! That kind of people don't need my bottles. What am I going to do now? My whole apartment will flood!'

This bizarre violation of the world order undercut the normal functioning of society. All its mechanical parts started falling apart. But the ideological part also disintegrated like rotten cloth. The children of official functionaries went to school with these pseudo janitors and false boilermen. They socialized and were friends: the officials' children often found it more interesting to spend time with these people, who had books and endless conversations. Their transparent, timeless existence seemed much more attractive than a regular career, a strict routine, money or material blessings. And the derangement of the children's minds began to influence the parent-officials as well.

Something was definitely rotten in our Denmark, from top to bottom. By the beginning of the 1980s everyone was expecting — well, we didn't know exactly what to expect: change, collapse, reforms, something. But when the first signs of the coming changes blew in, society stood stock still in astonishment, not believing its eyes and ears. After all, you can spend your whole life waiting for the Second Coming, but a 2000-year-old habit somewhat dulls the senses. What would you do if the archangel's trumpets actually began to sound right now?

*****

Here I must say something about one of the cardinal features of the old Russian intelligentsia, one which defined its state of mind for many long years, and which was partially responsible for its demise. For decades the intelligentsia was tormented by guilt with regard to the people. The 'people' is an even more abstract concept than the 'intelligentsia'. It's impossible to say whether any such creature even exists. It isn't the nation, it isn't an estate or a class. It isn't the peasantry as such; nor is it the workers; nor the merchant class. But these three classes together, it seems, constitute the people. It is certainly not the aristocracy. On the other hand, it isn't exactly the poor. Nor people of average means. But of course it isn't the rich either. If you are a peasant and poor, you have more chance of being called 'the people'. If you are a rich aristocrat, your chances are zero. Sometimes a rich merchant is considered the people, sometimes he isn't. It's not possible to define exactly who belongs to the people and who doesn't.

The only thing one can say definitively about 'the people' is that it is not the intelligentsia. The people is defined in opposition to the intelligentsia. There is an expression in Russian: to come out of the people. Usually it is applied to those who, coming from an uneducated family, receive an education and achieve something in life. But if you simply receive an education and don't achieve anything, then this expression can't be applied to you. On the other hand, there is another expression: to lose touch with the people. This indicates condemnation of a person who, having accomplished something, doesn't turn his thoughts toward the well-being of some imagined society. He who has lost touch with the people falls into the heresy of individualism. The people, meanwhile, is thought of as a mythological, indivisible whole. The intelligentsia is thought of as a collection of individuals, of personalities.

The intelligentsia actively participated in the creation and preservation of this myth, which takes various forms. In the soberest of them, members of the intelligentsia, horrified by the poverty and ignorance of the people — peasants for the most part, though not exclusively — tried to help as best they could (by opening schools and hospitals). In the dangerously romantic, impatient version, the intelligentsia urged the people to revolt, and was amazed when the people, always mistrustful and suspicious, turned them over to the authorities. In the Freudian-patriarchal version, exemplified by Lev Tolstoy (who formally was not a member of the intelligentsia but shared many of its myths), the people was the bearer of a certain unspoken wisdom and truth, a mystical grail, and there was no need to enlighten it or in any way impose one's own individual, sinful will upon it. Instead, one was supposed to bow down submissively before this higher truth and with the help of ritual actions, a kind of sympathetic magic, merge with the people.

This is what Tolstoy sought to do when he adopted the simple life: to undress, remove his shoes, lower himself, become impoverished, do simple, non-specialized work, reject the tools of civilization. The most important thing was to reject individuality, to merge with some mass. In Tolstoy's correspondence there is an astonishing document. A certain poor teacher wrote the great man a letter: knowing your generosity, I beg you to send me three rubles to get medical treatment for my sick daughter. The great man answers: no, he won't send three rubles, although, of course, it would be no trouble for him. He won't give money on principle, because many such poor, sick children are dying all over Russia, and although it's horrible, saving one child won't change anything; and on and on in the same vein. I read that correspondence twenty-five years ago and I've forgotten everything but that deadly letter, which stuck in my mind. The most awful thing, it seems to me now, is not that it was written, but that it was sent. The writer merged his voice with the voice of suffering humanity. He voluntarily joined in the collective suffering. But what about the father of the dying child? His humanity was his daughter, right then, concretely, at that very moment. I recall this story as one of the countless, typical examples of the Russian intelligentsia's thinking: instead of fixing the leaky roof above its own head, the Russian *intelligent* will wail about all the rains, past and future, that have ever fallen on the heads of Adam's descendants; and his or her torments, though abstract, are not false.

The intelligentsia continually tormented itself with guilt and for that reason attributed more positive qualities to the people than it actually possessed. It seemed to the intelligentsia that the people was steeped in a kind of special holiness, that in the communal principle there was a higher wisdom which it was not ours to judge, that the people possessed special, higher goals and its own special path to achievement. If peasants, acting out of anger and jealousy, burn down the house of a neighbour who somehow stands out or who has grown rich, they are actually enacting some higher judgment. It was said that as a whole, of course, the people was hardworking, sober and kind, and if in some cases it seems otherwise to us, then we are simply imagining it. At the same time any attempt to stand out, any desire to grow wealthy or succeed on a personal, individual basis, was unconditionally condemned.

The stubbornly idealistic idea remained that all one had to do was give the people freedom and a bit of help, direct the people and sacrifice oneself (the intelligentsia), and the people would suddenly reform, become enlightened and naturally turn to goodness, truth and selflessness — accept, in other words, the values of the intelligentsia. It was as if these values lay naturally at the base of all things, and only a lack of freedom prevented them from being seen. When the peasant, grown rich, became a greedy, rigid boss; when a man or woman, only just liberated, immediately began to oppress and exploit his or her fellows; when the little man, having received a little power, was transformed into the worst of tyrants — the intelligentsia wrung its hands in bewilderment, as if this was somehow a deviation from the norm. A wonderful trait, it might seem, to think better of a person than his actions warrant — but a dangerous one. To call on people (as they really did in fact) to rebel, and at the same time to say to them, 'We are the guilty ones', was a suicidal act. That is what happened. The intelligentsia's amazement that all the accumulated rage and aggression was directed toward it is even rather touching. How else could it have been?

* * * * *

Here it must be said that by the beginning of perestroika the intelligentsia had more or less lost its romantic passion for the people, and this was perhaps the most significant loss it sustained in over 150 years of existence. Too many insights and insults had accumulated for the intelligentsia to continue gazing amorously into the eyes of every hostile, crude, stubbornly angry, savage individual who might once have been automatically considered a representative of 'the people'. I think that the following may be suggested, though cautiously, as a matter of conjecture: the intelligentsia suddenly realized, quite late, that what it had admired all these years, what it had tried to understand and reconcile to itself, what frightened and mesmerized it, was fascism, protofascism, or at least the pre-condition for fascism.

There is still no institutionalized fascism in Russia. The regime that existed during the Soviet, particularly the Stalinist, period was called something different; and because of the Second World War and the victory, which was always referred to as the 'victory over fascism', it became particularly difficult to imagine and believe in the similarity between so-called communism and fascism. But this similarity was eventually noted and discussed, and in the last years before the fall of the Soviet Union it was almost the main topic of public debate. It was also the last splash, it seems, of Russian idealism. Finally the source of all evils was found: Communism, utopia. The paradox, obviously, was that by attacking utopia with all its might, the intelligentsia indulged in new utopian dreams: to rid itself of Communism in one fell swoop and wake up the next morning in a new society, perhaps not a prosperous, just society, but one moving full speed toward those goals.

The euphoria was particularly strong following the coup of August 1991. After three days of fear, the vigil at the White House (the building of the Russian parliament), the huge crowds, people coming out onto the streets to defend freedom (as the crowds themselves and the whole world supposed); after three ritual sacrifices (of young people who died under tanks); after the sudden and unexpected triumph of those who called themselves democrats — how could one not indulge in utopian dreams, not believe that everything would be fine now? If such terrifying days ended with the waving of palm branches, why not believe that at last everything had become clear to everyone (what was good and what was bad); why not suppose that the entire population would harmoniously set about building a new, peaceful tower of Babylon? On television they showed the same symbolic scenes: happy crowds, happy leaders waving from harmless tanks. A young man on roller skates with a Russian flag in his hands, swerving in and out of the crowds, slid over the square like an angel of peace and love.

* * * * *

The euphoria continued for about two months, then real life set in. The Russian democratic intelligentsia does a good job of talking, arguing, writing, advising, appealing, dreaming, creating rosy projects, protesting and pointing to where truth and good lie. But it doesn't know, realistically, how to resolve concrete problems. When dreaminess and idealism run into the crude prose of life, the intelligentsia is shocked. In this new historical phase, one of the worst problems has become the issue of ethnic conflict and ethnic or national identity. Here again, the tendency of the intelligentsia to romanticize other groups, including foreigners and ethnic minorities, has not helped it to understand what must be faced today.

Russians, more than any other people I know, are periodically seized by a passionate love for other peoples, nations, cultures, customs and languages — a passion which goes absolutely unrequited and is doomed to disappointment and misunderstanding. This passion has all the characteristics of romantic derangement. It is selective (Russians never went crazy over China, never dreamed about Portugal or Iran, are indifferent to Latin American culture, although Latin American writers are very popular); it is romantic, blind, demanding. As is often the case with young love, it chooses as the object of its passion a country that could not be more different from its own. Peter the Great fell in love with German and Dutch culture, although a happy marriage was clearly impossible. Peter, as is known, managed in a short time to make the Russian population dress in German clothes, he combed and shaved his subjects according to foreign fashion, made them drink, eat, dance and smoke like people in Europe. He founded the civil service bureaucracy, built a fleet, won a sea in battle so that the ships would have somewhere to float, and built an entire city — Saint Petersburg — so that he would not have to bother with reconstructing the old settlements, but could start with a clean slate. He dumped his Russian wife and married a German woman. He invited many Germans to Russia. He forced people to speak German and Dutch.

Peter died and everything began to cool gradually, the spark went out. And although all the Russian Tsars after Peter were German by blood, the love was forced, especially since the Germans did not appear to reciprocate. Russians were drawn in another direction: to France. That love affair began with Catherine the Great and Russian society was seized by the flame. At the beginning of the nineteenth century the Russian nobility spoke and read almost exclusively in French: some of them didn't even know Russian. They wore French clothes, ordered French outfits from Paris, ate and drank French food and wine. After the French Revolution crowds of French people hurried to Russia, fleeing from the guillotine, and were greeted with open arms. Many con-men and adventurists came as well, passing themselves off as aristocrats. They, too, were trustingly adored. Napoleon's march on Moscow didn't change much: the French were loved until the end of the nineteenth century, when the English were suddenly discovered.

Mass Anglomania was deterred by the events of 1917, but it had been gaining ground before the Revolution. Vladimir Nabokov's parents were still considered eccentric Anglomanes, but his generation (he was born in 1899) naturally inclined to Albion. At the very turn of the century there was a short but stormy burst of passion for Scandinavia. It was fashionable to read Ibsen and Hamsun, to play and listen to Edvard Grieg; it was fashionable to live in Finland in dachas built in the Scandinavian style, to take vacations in the north, amid the fir trees and waterfalls. Northern motifs became noticeable in poetry — dunes, sea gulls, flat seascapes, storms, the unsetting northern sun, swamps and romantic fishermen's daughters, awaiting the arrival of ships.

Love for Italy never faded, it merely shifted from love of ancient Rome to love of the Renaissance, and then to Italy in general, an imagined, blessed, sublime country where everything flowers and sings, as in the opera. Spasms of love for India could be detected — an even more enigmatic, mystical country, the source of wisdom, spirituality and all sorts of mysteries, revelations and philosophies. Tibet beckoned for the same reasons. Japan had all the elements for such love, but the Russo-Japanese war of 1905, which ended in a humiliating defeat for the Russians, interfered.

The general rule was: the more enigmatic the country, the greater chance it had to become the object of an unrequited passion. Thus, for example, clean, tidy Switzerland was never desired: people simply took refuge there to avoid political persecution. They sat out their time in coffee houses, eating fresh rolls and resolving the 'accursed' problems, the eternal Russian questions: What is to be done? Who is to blame? And how to distribute the rolls after seizing power? As far as I know, no country, no people responded to Russia with the reciprocity she naïvely expected. The most painful blow to Russian feelings came from the French. The Russian emigrés, fleeing to France after 1917, supposed that the French, sobbing, would embrace their Russian brothers and sisters, would take them in and help them in their time of trouble. We helped them after their revolution, and they will help us after ours, the Russians thought. Such naïvety, bordering on stupidity (or, if you prefer, such emotional purity and blind trust in the feelings of others) is possible only given an enormous love.

The last love affair was with America and everything American. It started long ago, during the war, when American humanitarian aid reached almost every Russian. In my childhood home in Saint Petersburg a strong, round, indestructible can which once contained American pork has stood for the last fifty years, filled with spare buttons. The Americans, whom Russians first saw at the end of the war, made their way into our souls, and the official anti-American propaganda of the cold war that soon began only strengthened our love. If the authorities assert something, then it isn't true — that is the usual Russian view of things. Russian Jews, emigrating in the 1970s, settled in America and wrote letters from there, exaggerating the virtues of the New World. There was no need to exaggerate: in the Russian imagination America was the distant Island of the Blessed, out there, beyond the ocean. Everyone was happy, rich, just. When perestroika began, the passion for America seemed to surpass all previous loves and was even mutual for a time. In any event, the Americans adored Gorbachev and seemed ready to love all Russians. Russians came in droves to America, to visit or to look for work, and for two or three years were so blinded that they not only failed to notice the most normal sorts of problems (and who doesn't have them?) but they couldn't even see the most obvious realities. Intelligent people said and wrote such exalted nonsense about everyday American life that one might think they had travelled to an enchanted, mythical country not found on any map. As Russians saw it, Americans were unusually honest, so much so that the idea of theft never even entered their heads; they were wondrously generous, kind and just; and of course there was no homelessness or unemployment in America and whoever denied this was either evil or envious. Americans liked hearing such pleasant things about themselves (who wouldn't?), and many believed that all this enthusiasm and praise was a sign that the Russians, finally freed from their horrible tyrants, had seen the light and would now try to become just as wonderful as the average American.

That part of the love affair with America was stormy and short. The partners soon cooled and separated, disappointed: they didn't understand one another. The Russian side imagined itself a Cinderella whom the handsome prince had finally noticed and appreciated. It counted on concluding such a marriage contract with the prince that it would never have to do anything again. Cinderella/Russia was so special, and had suffered so long in the dirt by the hearth, oppressed by her evil stepsisters and stepmother; now that her virtues had been justly recognized by a clean, wealthy prince, she could relax and rest. The prince, as it turned out, never had any intention of marrying the grubby simpleton: he simply welcomed her emancipation from the evil stepmother and offered to teach her how to set up democratic, equal relations with the evil sisters. He promised to give her a loan to get a new bucket and dustpan, and opened her eyes to the fact that all over the world people don't sit around waiting for a good fairy with glass slippers, but work hard, burn logs, collect the ashes and then sell them, and there's nothing wrong with that. One can easily imagine Cinderella's hurt feelings. If things are bad at home, and no one needs her in the outside world, then what is she to do?

*****

In the Russian mind a serious conflict has always existed between the law and the truth, the search for truth. At the worst moments of Russian history, disrespect for the law has played a positive role. For instance, during serfdom: by the law of the land a person had the right to be a slave owner, but according to the laws of conscience this was bad. Stealing is wrong according to the law, but if it is done in order to save your family from starvation, then it is good. To kill is of course wrong, according to both the law and conscience, but what about the murder of a tyrant and torturer?; and so on. How much good has been done in the world in violation of the law! But when society grows accustomed to viewing the very idea of the law with disdain, certain very important principles blur and the idea of a civil society is undermined.

During perestroika the most democratically oriented part of the intelligentsia nurtured the idea of remaking and rewriting all the laws, in order to establish new, just laws, which everyone would then agree to abide by, and life would begin anew, with a clean slate. But what happened during perestroika was that each group of the population, each people, each republic, each ethnic group — literally each and every one — started pulling the blankets over to its own side, so to speak. Everything needed to be changed, but everyone had different interests; and each, naturally, understood truth in his or her own way.

An important, popular group of democrats at that time started from the idea of giving everyone what he or she wanted, reckoning on a certain common conscience, a common system of values. Naïve, idealistic dreams. The intelligentsia of Russia, for instance, supported the independence of the Baltic countries and Georgia, figuring, strangely enough, that in acquiring independence these countries would love their Russian brothers tenderly for their aid and support. Everyone knows what actually happened. Discrimination against non-native ethnic groups in the Baltics is a fact that could have been foreseen, if one had looked at the situation soberly rather than shedding tears of brotherly love.

For many peoples of the former Soviet Union, the idea of 'Us' and 'Them' is very real, but for the Russian intelligentsia there is only 'Us'. With profound angst and bewilderment, the intelligentsia watched the cultural vandalism that resulted from the freedom it had supported. The Republic of Moldavia celebrated its freedom by removing from Kishinev a monument to Pushkin — Pushkin, our great poet, who was persecuted by the Tsarist regime and who was Ethiopian in origin! Estonia expelled the widow of the poet David Samoilov who had spent half his life in the Estonian city and sincerely believed that the Estonians loved him just as he loved them.

Thus the Russian intelligentsia at various times, against the advice of reason, wanted to merge into a unified 'Us' not only with 'the people', and with beloved, chosen foreigners, but also with the non-Russian ethnic groups that inhabit our country — although not with all of them. But everyone rejected the intelligentsia's amorous advances. They each, in one way or another, sent the intelligentsia an unequivocal message: come down out of the clouds and *do* something. Gorbachev's perestroika was the last such unambiguous signal the intelligentsia received: get to work.

* * * * *

Now we can return to the beginning of this essay. Perestroika was much harder for the intelligentsia than might have been expected. The intelligentsia thought that perestroika was designed for its benefit alone; but as it turned out, the intelligentsia was needed by no-one — not by the government, with which it wanted to work for perhaps the first time in its history; nor by the people, who acquired more and more fascist features right before our eyes; nor by foreigners who, as always, turned out to be not quite what we had imagined they were. The intelligentsia was left to itself, expected to do something, something small but concrete, in order to survive, not as a class — the intelligentsia isn't a class — and not as a kind of professional group in the population, for the intelligentsia transcends any group. But all of the intelligentsia's grand social plans for perestroika came to naught.

Instead a process of stratification of the intelligentsia began (and is still going on), in terms of its politics, economics and values. Before the August 1991 coup the intelligentsia's opinion on most matters was relatively united. Since everyone knew who the common enemy was, they thought of one another as friends. The attack on the Parliament in October 1993 can be seen as the real turning point: society was divided into people who believed that the Parliament had to be attacked; people who felt that it should not have been attacked under any circumstances; and people who still don't know what to think (I myself belong to the latter category). A second important event was the war in Chechnya, and the range of opinion was similar: it was necessary to go to war; we shouldn't have gone to war under any circumstances; and something had to be done, but not that way.

Stepping back from the broad spectrum of political views on Chechnya and on the actual political reality (about which we know almost nothing), it can be said that two traditional philosophies come into conflict here: the first approach requires that action be taken to ameliorate the situation, so as to avoid a subsequent regret that one stood by while everything all around was falling apart. The second views claims that it is better to do nothing at all if even a single drop of blood will be spilt. But a third possibility still exists, although it can hardly be called a point of view. Rather, it is the absence of a viewpoint; it is confusion, the aversion to taking sides, the need to escape. It is easy to judge or despise people who have no opinion on important questions, and this often happens. However, current bewilderment is in large part engendered by guilt — by a sense of moral responsibility for the strong preferences, opinions, political views that one held earlier — and I repeat that at present I include myself in this group. For example, at one time it seemed to me (as it did to many others I know) that the disintegration of the Soviet Union was a healthy, natural phenomenon. But when the collapse actually began — and it may never end, the process now looking like a chemical reaction in both directions — it turned out that so much blood was spilt, so many innocent lives were lost, that it was impossible not to be horrified and to start doubting. 'Now it's obvious that the disintegration of the country was a huge mistake', I heard recently from a person who played an active role in that process. I didn't take part, but I had an opinion, so the mistake is mine as well. Many people now feel that way; and if you find that your opinion, like a voice from the chorus, in effect approved someone's death, you feel reluctant to have opinions any more, you want to remain silent. You can't help remembering how the intelligentsia, supporting the revolution of 1917, ended up in the same position.

Political silence takes different forms. There is an exodus of the intelligentsia into commerce, and the exodus of the intelligentsia into emigration — temporary or permanent — continues. But emigration doesn't change much: all the new emigrants have blood ties to Russia, travel freely back and forth and often become more politically active abroad than they were at home. But the most productive type of escapism, if I may put it that way, is into art: painting, music and literature. Naturally, literature, the word, language, is closest to me. I am beginning to think that art is the only thing that truly belongs to the intelligentsia, the only thing it is good at, the only thing it does that is needed and valuable. Art is the only thing that is capable of reconciling and unifying. All the other methods have failed. It is the only real tradition — from the first literary works in the Russian language to those which appeared yesterday — running through all the Russian regimes. When Dostoevsky wrote his much quoted phrase 'Beauty will save the world', wasn't this what he really and primarily had in mind? Literature has tried to answer questions; over the course of two centuries, Russian literature has provided many answers, but none of them has been useful. The artist's job is not to give answers, but to ask questions, even questions which may not have answers. The artist's job is to search for beauty there, where he or she alone sees it; then, perhaps, others will open their eyes. In Russia, writers of all political persuasions have been persecuted; right and left have been jailed; monarchists and communists have been executed. In life no-one is saved; in art, everyone who created survived. So what does it matter what the answers were?

I believe that Russian literature is on the threshold of a renaissance. All the opportunities are there. When perestroika began, many naïvely thought that the changes in the political situation would automatically lead to the appearance of a new literature, that literature needs freedom. Others, seeing that this didn't happen, thought with equal naïvety that literature needs prohibitions, that the writer creates best in confinement, in prison conditions. Neither of these positions holds. The truth, it seems, is that neither prison nor freedom can prevent the artist from creating. He is independent of prohibitions and permissions; he is, as Pushkin said, his own highest judge. Now, when no one needs the artist, he is truly alone with himself and his own judge for the first time in Russian history. After all the political hopes, disappointments and shocks (and there will undoubtedly be more of them), the Russian intelligentsia is beginning to value art. The time of political art has passed and, as far as I'm concerned, we should be grateful for that. Art doesn't pay much and so it is less corrupt. Not many people need it and so it only attracts faithful adherents. The quantity of new literary journals and book stores, which somehow survive contemporary Russian conditions, violates the laws of statistics, and this is a very good sign. It's hard to predict what will come of this, but it's a very fruitful time for literature especially. Now, as never before, it is obvious: *ars longa, vita brevis.*

The Debate on Economic and Social Security in the Late Eighteenth Century: Lessons of a Road Not Taken

Emma Rothschild

'The heart of flint that has disgraced the beginning of the nineteenth century', William Godwin wrote in 1820, was the characteristic, in particular, of 'as many of us as studied the questions of political economy'. Political economy, he wrote in his extended response to Malthus's *Essay on Population*, is inimical to 'all the ramifications of social existence'; it sees the world as a cold and cruel scene, or as 'a city under the severe visitation of a pestilence' (Godwin, 1820: 110–12, 620). Like the poet Robert Southey, Godwin thought that the tendency of economists was to treat men in isolation from their social and public lives. 'Adam Smith's book is the code, or confession of faith of this system', Southey wrote in 1812.[1] 'Pluck the wings of his intellect, strip him of the down and plumage of his virtues, and behold in the brute, denuded, pitiable animal, the man of the manufacturing system!' (Southey, 1832: I, 111–12).

The point of this essay is to look at ideas of social development — including the social security and social integration of the poor — in the political economy of the late eighteenth century, and at their reflection in

I am grateful to Amartya Sen, Cynthia Hewitt de Alcántara, Stephen Martin and David Palfrey for helpful comments. A longer version of this paper appeared in *Population and Development Review*, December 1995.

1. In the essay, 'On the state of the poor, the principle of Mr. Malthus's essay on population, and the manufacturing system', published in his collected essays; see Southey (1832: I).

subsequent *laissez-faire* economics. The cruel reputation of political economy is quite undeserved, I will suggest, in relation to Adam Smith and to his most distinguished friends and followers in the period before the French Revolution. Social development, in their writings, was not inimical to but rather a condition for the development of commerce. The flint-hearted view of society, in which men and women are surrounded only by incentives, and inspired only by fear, was an innovation of the decade after Smith's death in 1790, and of the period of intense fright that followed the French Revolution.

In what follows, I will look first at Smith's own description of some of the constituents of social security and insecurity in the *Wealth of Nations*. I will then look at the development of these and related ideas in pre-Revolutionary France, and in particular at proposals of the great French statesman Turgot, and of the mathematician and economist Condorcet, for the reform of social assistance and for a social security insurance fund. These proposals were the object of intense criticism, it will be seen, in the period following the Revolution, and in discussions of the reform of the English Poor Laws; the rejection of social security was indeed of central importance to the quite different development of Smith's thought in Thomas Robert Malthus's *Essay on Population* of 1798. There were two sharply opposed views of social security in the *laissez faire* political economy of the last years of the eighteenth century, associated respectively with Condorcet and with Malthus. Malthus's views have been far more influential than Condorcet's in subsequent economic thought. But Condorcet's ideas — or the road which was not taken in the 1790s — are of continuing interest, it will be proposed, for modern economics.

In conclusion, I will suggest that Turgot's and Condorcet's ideas of social integration can illuminate modern debates over economic and social policy. The political economy of the late Enlightenment provides no support for the view of many contemporary proponents of *laissez faire* that social security is inimical to economic development, or that social equality is a form of luxury, to be promoted only in countries which are already rich. The characteristic presumption of Smith's early friends and followers in France was rather that political liberty, and the social integration of the poor, were causes (as well as consequences) of economic development. Smith and these friends were fierce critics of social institutions, including established religious and charitable foundations. But they were also concerned to invent new institutions, and new policies for social development. The debate over social institutions was indeed of central importance to the qualified optimism of the period immediately before the French Revolution. 'I too believe that humanity will win in the long run', Goethe wrote of Herder, from Naples in 1787: 'I am only afraid that at the same time the world will have turned into one huge hospital where everyone is everybody else's humane nurse' (Goethe, 1962: 312).

*****

The 'liberty and security of individuals' were, for Adam Smith, both the condition for the growth of commerce in early medieval Europe and its 'most important' consequence. The security was that of burghers, and especially of 'tradesmen and mechanics', who were thought of as 'a very poor, mean set of people'. They were subject to social discrimination: 'The lords despised the burghers, whom they considered not only as of a different order, but as a parcel of emancipated slaves, almost of a different species from themselves'. Tenants, too, were at the mercy of 'irregular and oppressive' taxes and compulsory services, and of the whims of their landlords, such that 'the security of thousands' might be 'endangered by the caprice or extravagance of one man' (Smith, 1976a: 284–5, 384–412).

The great transformation in European commerce came with the legal reforms of the feudal period: with what Smith's contemporary William Robertson described as the 'revolutions in property' which led to the rise of a 'spirit of industry', and to a revolution in the 'character and virtues of the human mind' (Robertson, 1769: 20, 36, 225). Smith's own 'great revolution' — 'a revolution of the greatest importance to the public happiness' — was a revolution in individual rights: the end of servitude, the right to own property, and the 'regular execution of justice'. Insecurity is, in Smith's description, inimical to industry, and in particular to the improvement of landed property. Security is, by contrast, the great object of individual endeavour. Even the short-sighted merchant, in Smith's famous metaphor of the invisible hand, is in quest of security: 'by preferring the support of domestic to that of foreign industry, he intends only his own security' (Smith, 1976a: 421–2, 456).

The characteristic of modern Europe, in Smith's description, and especially of modern England, is that liberty and security are to be extended to the poor and the landless. A civilized society is one in which even the poor have the right to secure lives. The security which was won so laboriously in medieval cities was the security of tradesmen and burghers. But Smith identifies individual security as the condition for industry among the labouring poor as well. It is not only yeomen who can be secure, independent and respectable: it is the 'equal and impartial administration of justice which renders the rights of the meanest British subject respectable to the greatest, and which ... gives the greatest and most effectual encouragement to every sort of industry' (ibid: 425, 610).

Smith was a fervent supporter of high wages, to take a first illustration, which he described as both the cause and the effect of national prosperity. He said of 'the liberal reward of labour' that 'as it is the necessary effect, so it is the natural symptom of increasing national wealth', and that 'to complain of it is to lament over the necessary effect and cause of the greatest public prosperity'. It was 'abundantly plain', he said, that an 'improvement in the circumstances of the lower ranks of the people' was of advantage 'to the society'. Such improvement was also a matter of social justice: 'no society can surely be flourishing and happy, of which the far greater part of the members are poor and miserable. It is but equity, besides, that they who feed, clothe and lodge the whole body of the people, should have such a share of the produce of their own labour as to be themselves tolerably well fed, clothed and lodged' (ibid: 91, 96, 99).

The 'liberal reward of labour' is for Smith an essential means of improving production. It 'increases the industry of the common people. The wages of labour are the encouragement of industry, which, like every other human quality, improves in proportion to the encouragement it receives'. He was entirely unconvinced by the proposition that people work harder when they are more afraid, or in years when real wages are low (which are 'generally among the common people years of sickness and mortality'). It 'seems not very probable', he said, 'that men in general should work better when they are ill fed than when they are well fed, when they are disheartened than when they are in good spirits, when they are frequently sick than when they are generally in good health' (ibid: 99–101).

Smith was well aware that he was questioning the received wisdom of contemporary employers, with regard to the invigorating effects of poverty. 'Masters of all sorts', he said, 'make better bargains with their servants in dear than in cheap years, and find them more humble and dependent in the former than in the latter. They naturally, therefore, commend the former as more favourable to industry'. He conceded that 'some workmen' will be idle for three days if they can earn their weekly wages with four days' work. But 'this, however, is by no means the case with the greater part'. A labourer is likely, rather, to be encouraged by the prospect of 'bettering his condition' — that is to say, of changing his position in society — and of 'ending his days perhaps in ease and plenty'; 'where wages are high, accordingly, we shall always find the workmen more active, diligent, and expeditious, than where they are low; in England, for example, than in Scotland'. Smith indeed describes the condition of Scottish women workers in pathetic terms. 'In most parts of Scotland, she is a good spinner who can earn twenty-pence a week'. 'Our great master manufacturers', meanwhile, 'endeavour to buy the work of the poor spinners as cheap as possible'; 'our spinners are poor people, women commonly, scattered about in all different parts of the country, without support or protection' (ibid: 99, 101, 134, 644).

It is interesting that Smith was even prepared to countenance government regulation in favour of workers: 'Whenever the legislature attempts to regulate the differences between masters and their workmen, its counsellors are always the masters. When the regulation, therefore, is in favour of the workmen, it is always just and equitable; but it is sometimes otherwise when in favour of the masters' (ibid: 157–8). Jean-Baptiste Say contrasted Smith's views explicitly, some years later, with the opinions of master employers. 'One meets leaders of industry', he said, 'who, always ready to find arguments to justify the consequences of their greed, maintain that the worker who is better paid works less, and that it is good that he should be stimulated by need. Smith, who had seen a great deal and was a perfectly good observer, is not of their opinion'. 'The comfort of the inferior classes is in no way incompatible ... with the existence of the body social', Say added, paraphrasing Smith: 'a shoemaker can make shoes just as well in a heated room, dressed in a good suit, when he is well-fed and feeds his children well, as when he works freezing in the cold, in a hovel, in the corner of the street ... The rich should therefore abandon this childish fear of being less well-served, if the poor man acquires comfort' (Say, 1972: 385, 474).

Smith's description of the social context of consumption provides a second illustration of his view of social development. He is no more concerned by the supposed frivolity of the poor than by their supposed indolence. He is quite undisturbed, for example, by the desire of workers to have several days of 'relaxation' in each week, which he describes as often the consequence not of indolence but of 'over-work': 'excessive application during four days of the week, is frequently the real cause of the idleness of the other three, so much and so loudly complained of'. He is not even averse to occasional dissipation: 'great labour', he says, 'requires to be relieved by some indulgence, sometimes of ease only, but sometimes too of dissipation and diversion'. He is struck, however, by the *lack* of dissipation in the consumption of the poor. He contrasts the 'disorders which generally prevail in the economy of the rich' with the 'strict frugality and parsimonious attention of the poor'. The common people, he says, are in general far more 'strict or austere' than 'what are called people of fashion'. His principal examples of 'indolence' are landlords, and the established clergy (Smith, 1976a: 98, 100, 265, 789, 794).

Smith describes the consumption of the poor, in a famous passage, as the means to a specifically social end: the end of decency in society, or of having a creditable position in public life. He defines 'necessaries', in his account of indirect taxation, as those commodities which 'the custom of the country renders it indecent for creditable people, even of the lowest order, to be without'. The labouring poor are seen as prudent, reflective, civic beings, concerned for their public position and subject in particular to the emotion of shame: 'a creditable day-labourer would be ashamed to appear in public without a linen shirt'. These civic emotions are common, interestingly enough, to men and women alike. Leather shoes are for example necessities in England: 'the poorest creditable person of either sex would be ashamed to appear in public without them'. In Scotland, they are necessities only for men of the lowest order; 'but not to the same order of women, who may, without any discredit, walk about bare-footed'; 'in France, they are necessaries neither to men nor to women' (ibid: 869–70).

Consumption is in general, for Smith, a means to the end of social integration, and social renown. 'To what purpose is all the toil and bustle of this world?', he asks in his *Theory of Moral Sentiments*; 'what is the end of avarice and ambition, of the pursuit of wealth, of power, and pre-eminence?'. His answer is that people are concerned, above all, with their positions in society: 'to be observed, to be attended to, to be taken notice of with sympathy, complacency, and approbation, are all the advantages which we can propose to derive from it'. The dismal destiny of the poor consists in being looked at without sympathy, or in not being looked at at all, to be 'out of the sight of mankind' (Smith, 1976b: 50–51).

'A man of low condition', Smith says in the *Wealth of Nations*, 'is far from being a distinguished member of any great society'. When 'he remains in a country village', he is at least 'attended to'. 'But as soon as he comes into a great city, he is sunk in obscurity and darkness. His conduct is observed and attended to by nobody'. Smith is willing, here too, to countenance the intervention of government in the interests of the social integration of the poor. He thus proposes to enliven the lives of people in great cities — for whom 'respectable society' is often to be found only in small sects, whose 'morals' are 'rather disagreeably rigorous and unsocial' — as a matter of public policy: by support for 'the study of science and philosophy', and by 'the frequency and gaiety of public diversions'. He is strongly opposed to 'direct taxes upon the wages of labour', which he describes as 'absurd and destructive', and also to 'a tax upon the necessaries of life'. But he favours taxes on luxuries, and especially on the luxuries of the rich. He is in favour, for example, of progressive tolls on 'carriages of luxury' ('somewhat higher in proportion to their weight'), such that 'the indolence and vanity of the rich is made to contribute in a very easy manner to the relief of the poor' (Smith, 1976a: 725, 795–6, 865, 871).

Smith's account of public instruction, thirdly, is a further eulogy to the social integration of the poor. It is not enough that the poor should be able to appear in public without shame; they should also be able to take part without shame in public and political discussion. The budgets of the poor are generally prudent, in his description; he speaks of the labourer who works hard in the hope of ending his days in ease, or of the 'labouring poor' who are impeded by unjust taxes in their ability 'to educate and bring up their children' (ibid: 508). But he sees an essential role for government in providing free or subsidized education for 'the children of the common people'. He is insistent, from the beginning of the *Wealth of Nations*, on the equality of natural talents. The difference between the philosopher and the common street porter, he says, 'seems to arise not so much from nature, as from habit, custom and education'. Their 'very different genius' is the consequence of the division of labour, more than its cause. People are at first 'very much alike'. They are not born 'stupid and ignorant', but are made so by their 'ordinary employments'; by the simple, uniform nature of the work they can get, and by the circumstance that their parents, 'who can scarce afford to maintain them even in infancy', send them out to work as soon as they can (ibid: 28–9, 782–5).

The public 'can facilitate, can encourage, and can even impose' a system of education on 'almost the whole body of the people', Smith says. The 'most essential parts of education' are 'to read, write and account', and even the poorest people should 'have time to acquire them' before they begin their working life (ibid: 785–8; see also Skinner, 1993). Smith is resolute in identifying education as something which is good in itself, and not as the means to a distinct, commercial end. When he does talk of universal instruction as a means, it is in relation to the political ends of the society, or to the common interest in political security. People 'of the inferior ranks' who are instructed are 'more disposed to examine, and more capable of seeing through the interested complaints of faction'; they are less susceptible to 'wanton or unnecessary opposition to the measures of government'. This is the Enlightenment idyll, of universal public discussion among thoughtful, reflecting, self-respecting individuals. It is also Smith's own particular idyll, of reciprocal respectability. People who are instructed, he says, 'feel themselves, each individually, more respectable, and more likely to obtain the respect of their lawful superiors, and they are therefore more disposed to respect those superiors' (ibid: 788). Even parents, he recommends in the *Theory of Moral Sentiments*, should treat their children with respect, since 'respect for you must always impose a very useful restraint upon their conduct; and respect for them may frequently impose no useless restraint upon your own' (Smith, 1976b: 220–22).

*****

Smith's ideas of social and economic security were strikingly close to those of his great French contemporary Turgot — of whom he wrote that he was 'a person whom I remember with so much veneration', whose policies 'did so much honour to their Author and . . . would have proved so beneficial to his country' — and Turgot's reforms of the 1770s constituted the first major political experiment in these ideas (Mossner and Ross, 1987: 286). For Turgot, as for Smith, the two principal objectives of economic reform were to end restrictions on free trade in subsistence food, and restrictions on industry imposed by guilds, corporations and apprenticeship regulations. 'The unlimited, unrestrained freedom of the corn trade' is the best preventative of scarcity, Smith wrote in 1776, and the best policy 'for the people'; for Turgot, a few years earlier, 'freedom is the only possible preservative against scarcity' (Turgot, 1913–23, III: 267; see also Rothschild, 1992a). Smith proposed to 'break down the exclusive privileges of corporations, and repeal the statute of apprenticeship, both of which are real encroachments upon natural liberty'; for Turgot, 'the destruction of the mastership guilds', with the 'total freeing' of the poor from corporate restrictions, was as significant as the reform of the corn trade, and 'will be for industry what the former will be for agriculture' (Smith, 1976a: 470; Turgot, 1913–23, V: 159).

Turgot's objective, as a provincial administrator and later as Minister of Finance of France from 1774 to 1776, was to try to introduce 'complete freedom' in agriculture and industry. But the process of reform was turbulent, as he discovered, and especially so in a country where people were still poor and insecure. Smith wrote the *Wealth of Nations* in the course of the

1760s and 1770s, at the end of a period of prodigious growth in the English economy, during which England came to surpass Holland as the emblem of economic modernity in Europe, and in which the standard of living of the English poor increased substantially; in E. A. Wrigley's words, 'real wages were probably rising from the mid-seventeenth century until about 1780' (Wrigley, 1987: 235). In France, by contrast, people in the poorest regions were still vulnerable, as late as the 1770s, to the intense insecurity of impending scarcity.

Turgot was himself 'Intendant' of the Limousin region during one of the last subsistence crises in eighteenth century France, and the experience of the crisis exercised a profound influence on his subsequent policies. Food prices increased sharply in the Limousin in 1769–70, following a sequence of bad harvests, and mortality began to increase, especially in remote rural areas. The freedom of the corn trade could not prevent scarcity 'in the first years when it is established', Turgot concluded. 'If commerce is to be able to prevent scarcities entirely', he wrote to Dupont de Nemours, 'the people would already have to be rich'. The prospects of the landless poor were evidently insecure. The margin of the 'superfluous' is for the poor 'very necessary', Turgot wrote; it provides the possibility of 'some small enjoyments', or 'of a small fund which becomes their resource in unforeseen cases of illness, of rising prices, of being out of work'. But in the crisis of 1770, 'the people have only been able to survive by using up all their resources, by selling, at very low prices, their furniture and even their clothes'.

The relief of the poor in France was based, in general, on individual charity or on religious institutions; on parish charity in the countryside, and on large hospitals or 'foundations' in the cities. The charity of individuals (or their 'moral economy') provided insufficient security in the crisis of 1770. There was a tendency for prosperous farmers to send away their share-croppers, Turgot wrote, and to 'turn out their domestics and servants'; 'the purely voluntary submissions' of the well-off, he determined, should be augmented in certain parishes by a 'roll' of contributions, proportionate to the contributor's means. He also became aware of the fragility of the system of parochial relief. He directed his officials, for example, to distribute copies of his instructions to individual landowners in each parish; 'this attention will be particularly necessary in those parishes where you know that the local priest, either by lack of capacity, or by some vice of his character, or simply because he does not have the confidence of his inhabitants, cannot manage the operation on his own and make it succeed' (Turgot, 1913–23: III, 223, 234, 267, 288, 347, 358, 384).

The large hospital foundations had been the object of Turgot's bitter criticism as early as 1757. They were places of 'vanity, envy, hatred', he wrote (in an article in d'Alembert and Diderot's *Encyclopédie*), where the wardens went from patient to patient, 'mechanically and without interest', distributing food and remedies 'sometimes with a murderous negligence'. They were to be contrasted, in particular, with the 'free associations' or 'societies' of citizens for voluntary support of those in need, of which 'England, Scotland and Ireland are full': 'what happens in England can happen in France, and the English, whatever one might say, do not have the exclusive right of being citizens' (ibid: I, 587, 592).

When Turgot himself was Minister of Finance, he initiated a major reform of relief and welfare policies. His principal strategy, in the Limousin, had been to provide short-term employment in public works, and he attempted to generalize the policies in other regions. He established a system of 'Charity Offices and Workshops', on the grounds that the poor who are able to work 'need wages, and the best placed and most useful alms consist of providing them with the means of earning'. He laid special emphasis on 'the employment of women', which he described as 'an objective no less worthy of attention' than the employment of men; he proposed that the Charity Offices should advance spinning wheels to 'poor women', and should pay for instruction in spinning in each village. He insisted, as Minister of Finance, on providing income for women and children as well as for male labourers, since it was they who suffered most in periods of scarcity; 'it is this part of the family for whom one must find employment and wages', and the wages should be 'distributed to all consumers, even to the children of whom the family is constituted' (ibid: III, 125–6, 214–15, 225, 250; IV, 499–503).

Turgot's other major reform of relief policies consisted of efforts to reduce the numbers of people who were unemployed for long periods. France was in the sway, in the 1760s and 1770s, of one of the Ancien Régime's periodic preoccupations with the problem of conspicuous indigency. The notorious 'Depots of Mendicity', or workhouse-prisons, had been established in the 1760s; as a student in Strasbourg, Goethe followed Marie Antoinette's progress towards Paris as a young bride, and observed with some irony that 'before the Queen's arrival, the very reasonable regulation had been made that no deformed persons, no cripples nor disgusting invalids, should show themselves on her route' (Oxenford, 1974: I, 396). Begging, in the words of a famous survey of 1779, 'turns Society into ... a monstrous collection of enemies who know only how to fear, to hate, to avoid and to harm one another' (Académie des Sciences, 1779: v).

Turgot's policy, as Minister, was to close down the Depots of Mendicity (which were run by unscrupulous private 'entrepreneurs' who shaved all the inmates' heads and speculated on subsistence food) (Bloch, 1908: 168–78). 'His Majesty's intention is that you should immediately release all those who are confined', Turgot wrote in 1775 to the Intendant in Normandy; the only exception was to be made for those who are 'absolutely dangerous and incorrigible'. All others were to be sent home, with travelling expenses, and with a pension for those who could not earn their own living. A few were suitable to enroll in special companies of 'soldier workers', where they would be trained and would also be remunerated with a share of their 'net product'. These people, Turgot said firmly to the Intendant, should be provided with shoes, and should not be accompanied on their way by marshals, since 'they are destined to be free' (Turgot, 1913–23: IV, 515–24; V, 426).

Turgot's strategy, in general, was to provide political guidelines, and public finance, for short-term income security. In his elaborate plan of local government — his *Mémoire sur les Municipalités* of 1775 — one of the roles of the proposed 'municipal assemblies' was to be the overseeing of support for the poor. There would be 'a general plan given by government', but the implementation of the policies against poverty would be in the hands of local assemblies. It is interesting that Turgot envisaged quite extensive public participation in the discussion of these policies. The assemblies would in general be elected only by people who owned property in land (and who were thus liable for direct taxes). But in their functions with respect to relief, the assemblies should also take account of the 'views of people without landed property'; Dupont de Nemours recounts that Turgot made clear in his pencil emendations to a draft of the local government plan that such consultation was essential in matters 'which can affect the freedom of individuals' (Turgot, 1913–23: IV, 591; and comments of Dupont de Nemours: 571; see also Condorcet, 1847–9: V, 140).

Turgot 'tried to provide a rational basis for assistance, and to impose upon it a reflective practice', the historian of French social policy Camille Bloch wrote in 1908. His achievement was to 'orient assistance clearly towards a public, official organization'; 'he wanted it to favour the dignity of the individual'. His principal innovations, like the free market reforms which were so admired by Adam Smith, barely survived his own period of office. Even the Depots of Mendicity were re-established.[2] But the cumulative effect of Turgot's policies, over the 'last years of the Ancien Régime', was to promote 'the movement in favour of local home-based assistance, and assistance through the provision of employment' (Bloch, 1908: 182, 208, 210, 443).

Condorcet was Turgot's biographer and close friend, and in the years following Turgot's fall he developed a quite general theory of social security in periods of economic transformation. The 'quantity of happiness', he wrote in his *Reflections on the Corn Trade*, published in 1776 in support of Turgot's reforms, is not a proper object of government policy. But 'welfare' (*bien-être*) is a necessary (although not sufficient) condition for happiness. Condorcet defines welfare in the minimal sense 'of not being exposed to misery, to humiliation, to oppression'. It is in this sense a proper government objective, or a 'duty of justice': 'it is this welfare which governments owe to the people'. 'That all members of society should have an assured subsistence each season, each year, and wherever they live; that those who live on their wages, above

---

2. The villainous M. Valenod in *Le rouge et le noir*, 'a chronicle of 1830', is thus director of the Depot in Verrières, where he is disturbed during a dinner party by an inmate singing (Stendhal, 1990: 168–9).

all, should be able to buy the subsistence they need: this is the general interest of every nation' (Condorcet, 1847–9: XI, 111, 135).

Like Turgot and like Smith, Condorcet believed that the causes of indigence and misery were to be found in 'bad institutions'. In his essay of 1788 on Provincial Assemblies, which was a continuation of Turgot's work on Municipalities, he sought to identify the 'causes of poverty', which included the lack of 'general competition', 'bad laws in relation to the corn trade', 'the spirit of regulation', and the 'chains' with which commerce is encumbered. The most efficient policies are those whose effect is to prevent people from becoming poor, as distinct from supporting them in 'public establishments': 'Calculate how much the Poor Rate, in England, has cost for supplying their consumption, and see what an enormous difference there would be in the effects if this same capital had been employed in industry' (ibid: VIII, 454, 456, 458–59; XII, 648).

One outcome of Turgot's and Condorcet's ideas was to be found in the reform of social assistance in the early years of the Revolution. The 'Committee on Mendicity' of the Constituent Assembly, led by another of Turgot's younger associates, the Duc de la Rochefoucauld-Liancourt, described assistance as a matter of justice, rather than compassion. It could not be organized on a purely local basis ('as in England'); 'whenever there exists a class of men without the means of subsistence, then there is a violation of human rights; then the social equilibrium is broken'. Nor could it be left to the Church, or to individual foundations. The state should therefore assume both the responsibilities and the rights (including the property) of religious incorporations. Liancourt's early twentieth century biographer describes a liberation of individuals and of the state: 'In the face of the corporations, "fictions and moral persons", there arose the secularized state, enfranchised at the same time as the individual from subjections and privileges'. The duties of the state, or of society, were meanwhile to be substantially more extensive than those of the old institutions: 'it is no doubt an imperious duty of society to assist poverty, but the duty of preventing it is no less sacred and no less necessary' (Ferdinand-Dreyfus, 1903: 172–5; Bloch, 1908: 443).

Condorcet himself proposed two principal policies to prevent poverty. The first, which he identified repeatedly with Smith, was for universal public instruction. Of all the causes of poverty, he said, there was only one which was the result of economic progress rather than of bad institutions. This was the poverty which follows from the 'invention of machines', leading to unemployment among 'workers who only know how to do one thing'. But if workers were better instructed, then this cause of poverty would be transitory. The enormous scheme of general competitive equilibrium would only work efficiently if people could move from one industry and one employment to another: 'It will no doubt take some time to reestablish equilibrium, but the time will be shorter to the extent that there is greater freedom'.

Like Turgot, Condorcet was insistent that women as well as men should be educated, and not only trained but retrained: 'We are proposing an education which is common to men and to women, because we do not see any reason for it to be different'. Public instruction was moreover of political as well as economic importance. People who did not know how to count, or who did not understand local laws, were dependent on others: 'social institutions must combat, as much as is possible, this inequality which produces dependency'. Instruction was necessary, above all, to 'make a reality out of the enjoyment of the rights which are assured to citizens by law'; 'does a being enjoy his rights, when he is ignorant of them, when he cannot know if they are being attacked?' (Condorcet, 1847–9: VIII, 458–9, 474–6; see also VII, 452).

Condorcet's second policy to reduce poverty, and to promote social equality, was the direct outcome of his and Turgot's experience with the subsistence crises of the 1770s. One cause of poverty, he said, was poverty itself: 'every family which has neither landed property, nor moveable property, nor capital, is exposed to fall into misery at the smallest accident. Thus, the more families there are who are deprived of these resources, the larger the number of the poor'. His proposal, in these circumstances, was for a system of social savings banks: 'it is by opening savings banks, by means of which small savings can ensure help in illness and in old age, that one can prevent misery' (ibid: VIII, 453, 461).

In the course of the Revolution, Condorcet developed his ideas for the security of small savings into a major system of social security. He described the dreadful prospects for women whose existence 'depends absolutely on the length of the husband's life', and 'the great number of invalids, old people, women and children who fall from a state of comfort to a state of poverty and misery'. It was essential, he said, for a man whose subsistence depended on his labour to be able 'to ensure by his savings the means of subsistence in old age', as well as resources for his wife and children in case of his sickness or death. This required that there should be secure means of placing very small — even daily — savings. Such means could be provided, he said, by 'associations of individuals, or by companies, or even by the state'. But associations would be too small to provide national benefits, and private companies would find the business insufficiently profitable.

The state could by contrast set up social insurance establishments on a scale such as to take advantage of 'tables of general mortality' for very large numbers of insured individuals. Associations and private companies, Condorcet said, had the option of only admitting 'people of whom a doctor, whom they trusted, would state that they could reach the mean life expectancy for people of their age; but this means could not be appropriate for an establishment created for the public'. Public establishments would have the further advantage of helping to reduce the national debt. They would in general, by increasing 'the number of people whose lot is secured', help to bring about a different sort of society: 'something which has never before existed anywhere, a rich, active, populous nation, without the existence of a poor and corrupted class' (ibid: XI, 389, 392–4, 397, 402; see also Baker, 1975: 280–82).

Condorcet's establishments for social security became, at the very end of his life, the foundation of his ideal of future progress. He speaks, in the *Esquisse des Progrès* that he wrote while in hiding from the Jacobin Terror, of the great enterprise of 'using chance to oppose chance itself'. People will in the future be secure in their old age, with their savings augmented by those of others who died before retirement. Families will have some 'compensation' if they are afflicted by the premature death of a father. The insurance establishments 'could be formed in the name of social authority', but they could also, Condorcet now concluded, be 'the result of associations of individuals', since the principles of social insurance would be more familiar. 'The application of calculation to the probabilities of life expectancy, and to financial placements' will be used, henceforth, in the interest of 'the entire mass of society'. The coming epoch would be one not of 'entire' economic equality — which Condorcet considered to be inimical to industry — but of the 'social equality' which would follow from instruction and social insurance; 'social equality is sufficient, virtually on its own, to destroy two principal causes of corruption and prejudices, which are indolence and bad example' (ibid: VI, 247–8, 591).

*****

Condorcet's ideal of social security was the subject of very little practical or political interest after his death in the Revolutionary Terror. But it played an indirect role, in the course of the 1790s, in the subsequent counter-revolution of political economy. Malthus qualifies his 1798 *Essay on the Principle of Population*, on the title page of the first edition, with the subtitle 'as it affects the Future Improvement of Society with Remarks on the Speculations of Mr. Godwin, M. Condorcet, and other Writers', and his denunciation of Condorcet begins with establishments for social protection. Condorcet, Malthus notes, 'proposes that a fund should be established, which should assure to the old an assistance, produced, in part, by their own former savings ... [T]he same, or a similar fund, should give assistance to women and children, who lose their husbands, or fathers'. 'These establishments ... might be made in the name, and under the protection, of the society. Going still further, he says that by the just application of calculations, means might be found of more completely preserving a state of equality, by preventing credit from being the exclusive privilege of great fortunes, and yet giving it a basis equally solid, and by rendering the progress of industry, and the activity of commerce, less dependent on great capitalists'.

Malthus is full of derision for what he describes as Condorcet's 'enchanting picture': 'such establishments and calculations may appear very promising upon paper, but when applied to real life, they will be found to be absolutely nugatory'. One reason is the softening of 'the goad of necessity': 'If by establishments of this kind, this spur to industry be removed ... can we expect to see men exert that animated activity in bettering their condition, which now forms the master spring of public prosperity?'. The other reason is the principle of population itself. 'Were every man sure of a comfortable provision for a family, almost every man would have one; and were the rising generation free from the "killing frost" of misery, population must rapidly increase'(Malthus, 1986: I, 7, 55–6).

Malthus's *Essay* played a critical role in the reconstruction of political economy in the decades after Smith's death. Smith himself, Malthus writes, makes a 'probable error' in mixing two distinct inquiries: into 'the wealth of nations' and into 'the happiness and comfort of the lower orders of society'. In his *Principles of Political Economy*, Malthus identifies more of Smith's infelicities, including using 'exceptionable' language about landlords and speaking of them 'rather invidiously, as loving to reap where they have not sown'; underestimating 'the production contributed by the capitalist'; and talking about humanity — 'if humanity could have successfully interfered, it ought to have interfered long before', Malthus concluded, 'but unfortunately, common humanity cannot alter the funds for the maintenance of labour' (ibid: I, 107; V, 63–4, 181). Smith's supposed indulgence of the poor, and of their interest in a secure existence in society, was indeed a source of recurring irritation to his early critics. Edmund Burke, for example, wrote in 1795 that 'nothing can be so base and so wicked as the political canting language, "the laboring poor" '. This is a phrase which Smith uses repeatedly in the *Wealth of Nations* (and ten times in the few pages of his discussion of wages). Providence sometimes withholds 'necessaries' from the poor, Burke says; 'it is not in breaking the laws of commerce, which are the laws of Nature, and consequently the laws of God, that Divine displeasure is to be softened' (Burke, 1894: V, 135; see also Rothschild, 1992b).

One of Smith's oddest critics, William Playfair, chides Smith explicitly for ignoring the useful goading of the poor. In a footnote added to Smith's chapter on wages, reproduced in Playfair's annotated 'eleventh edition' of the *Wealth of Nations* (of which the *Edinburgh Review* wrote that 'in the whole course of our literary inquisition, we have not met with an instance so discreditable to the English press'; Horner, 1806: 471[3]), Playfair writes that 'Mr. Smith in this case, as well as in that of bearing increased taxation, puts down nothing for that great spring of industry — necessity'. 'Mr. Smith sets nothing down for Necessity, the nurse of Industry', he repeats in a later footnote; the negative effects of high taxes are 'counterbalanced in a great degree by the spur they are to industry'. Even education, for Playfair, is a distraction from industry, or 'business'. He thus objects, in his 'supplementary

---

3. For the attribution to Horner, see Fetter (1957: 9).

chapter' on education, to Smith's concern with the instruction of the poor: 'Whether or not it contributes to the comfort and happiness of the working man, to read and write, is a question not necessary to decide, and probably not very easy ... Reading frequently leads to discontent, an ill-founded ambition, and a neglect of business ...' (Playfair, 1805: I, 131; II, 27–8; III, 243).

Malthus's first *Essay* was a work of polemic, written in one of the periods of greatest trepidation over the effects of the French Revolution on the English poor.[4] But its influence on the subsequent interpretation of political economy, and especially on ideas of economic and social security, was very much greater than that of Malthus's later, more reflective writings on population, or of his own *Principles of Political Economy*. In the controversy over the successive *Essays*, even the *Wealth of Nations* was read in the light of Malthus's theory. Some critics continued to distinguish between Smith and his followers. Godwin, for example, quoted Smith's eulogy to high wages in his answer to Malthus, and said that 'it is refreshing to come to such sentiments as are here put down, after the perusal of such a book as that of Mr. Malthus' (Godwin, 1820: 611). Sismondi complained in 1819 that 'in England, the disciples of Adam Smith have distanced themselves from his doctrine', whereas Smith himself 'considered political economy as an empirical science; he made the effort to examine every fact in its social context' (Sismondi, 1819: I, 57). But for many early nineteenth century critics of political economy, as for Southey, Smith was little more than the precursor of Malthusian reform.

The idea of social security, or of the social context of individual enterprise, was of central importance to the controversy between Malthus and his critics. Malthus himself was convinced, quite generally, of the beneficial effects of fear: 'if no man could hope to rise, or fear to fall, in society; if industry did not bring with it its reward, and idleness its punishment, the middle parts would certainly not be what they now are' (Malthus, 1986: I, 129). His view of human nature (or at least of the human nature of the poor) is dispiriting. We must 'consider man as he really is, inert, sluggish, and averse from labour, unless compelled by necessity', he writes in the first *Essay*; the tendency of leisure, 'taking man, as he is', is to 'produce evil rather than good'; 'the general tendency of an uniform course of prosperity is, rather to degrade, than exalt the character'. The English Poor Laws had 'powerfully contributed to generate that carelessness and want of frugality observable among the poor ... [T]he labouring poor, to use a vulgar expression, seem always to live from hand to mouth'. The danger of Condorcet's social establishments, Malthus notes in the first *Essay*, was that they

---

4. It was written, too, under the intermittent effect of 'a very bad fit of the toothache', of which Malthus himself said that he was scarcely unaware even 'in the eagerness of composition' (Malthus, 1986, I: 81).

would free the poor from the frost of misery; in the *Essay* of 1826, it was that the poor would be 'free from the fear of poverty' (ibid: I, 127, 129, 130; III, 321, 366).

For Malthus's critics, by contrast, incentives for hard work were to be found in hope (or in greed) as much as in fear. Is the 'primary source of evil' to be found 'in human institutions, or in the laws of nature', Southey asked, and are 'lust and hunger' to be seen as 'independent of the reason and the will'? (Southey, 1832: 78, 81, 86, 111). It is not 'the Law of Nature', for Godwin, which is the cause 'of mischief to society', but 'the Law of very artificial life' (Godwin, 1820: 20, 601–2). Malthus 'has always a certain quantity of misery *in bank*', William Hazlitt wrote: 'so many poor devils standing on the brink of wretchedness, as a sort of out-guard or forlorn hope, to ward off the evils of population from the society at large'. Condorcet's social fund would have provided either temporary assistance, or assistance to 'a surviving family, in case of accidents', Hazlitt says: 'did Mr. Malthus never hear of any distress produced in this way, but in consequence of the idleness and negligence of the deceased?'. Hazlitt, like Condorcet, thinks that the security of the poor is most unlikely to have 'degraded the human character'; 'if the English poor laws are formed upon this principle [of security], I should, I confess, be very sorry to see them abolished' (Hazlitt, 1807: 236–7, 242, 262).

The English Poor Laws were for Malthus the source of 'poverty and wretchedness'. He compared England unfavourably with Germany, for example, of which no part 'is sufficiently rich to support an extensive system of parochial relief'. 'From the absence of it', he said, the lower classes are in some parts 'in a better situation' than the English poor. Holland's ability to associate commercial prosperity with social security he attributes, quite oddly, to trade, emigration, and 'extreme unhealthiness' (with consequent high mortality): 'these, I conceive, were the unobserved causes which principally contributed to render Holland so famous for the management of her poor, and able to employ and support all who applied for relief' (Malthus, 1986: III, 522). The Irish critic George Ensor summarized Malthus's account quite starkly: it was 'disease and rapid mortality' which enabled Holland 'to support its poor with distinguished facility' (Ensor, 1818: 259).

Ensor's own account, by contrast, was that the great advantage of Holland consisted in its political institutions: 'the fact is, Holland was a republic', its people were industrious, and 'considering its territory pre-eminently opulent … this opulence was transmitted down to the lowest orders of the people'. Idleness and negligence were more generally the effect, as much as the cause of social evils. 'Why should the Irish be industrious, who will not receive the profit of their industry?', Ensor asked. The English complain in India, he says, citing Orme's *History*, that ' "the people are without industry, and without energy" '. But indolence can itself be a counsel of prudence. Ensor quotes Orme's own observation, 'concerning these very people aërially predisposed to indolence, "that a dread of extortion or violence from the officers of the district makes it prudent in him to appear and *to be poor*" ' (ibid: 396–7, 402–4, 495–6, 501).

The profound distinction, in these controversies, is over the effects of insecurity; over whether it is fear or hope which drives the lives of the poor and the rich. One of Malthus's inconsistencies, as Hazlitt observed, is to presume different incentives, or checks, for different groups of people. If the 'fear of misery' were to be the check to population 'among the rich', Hazlitt said, then 'the world would be one great work-house', with no 'room for such a number of poor gentlemen' (Hazlitt, 1807: 261). It is only the poor, by implication, who must be goaded by necessity; and only the rich who are to be goaded by hope. But for Smith, as has been seen, the rich and the poor are 'very much alike'. Hope, or the restless desire of bettering one's condition, is a universal inducement to industry. Fear is the inducement to universal misfortune. In Smith's words, 'fear is in almost all cases a wretched instrument of government, and ought in particular never to be employed against any order of men who have the smallest pretensions to independency. To attempt to terrify them, serves only to irritate their bad humour' (Smith, 1976a: 798). For Condorcet, in his essay on monopoly of 1775, 'fear is the origin of almost all human stupidities, and above all of political stupidities ... [I]n curing men of fear, one would cure them of many prejudices and many ills' (Condorcet, 1847–9: XI, 54).

＊＊＊＊＊

Condorcet's establishments for social insurance and Malthus's Poor Law reforms constitute the two opposite destinies of late eighteenth century political economy. Both Condorcet and Malthus presented their proposals as in the spirit of Smith and of the *Wealth of Nations*, and both believed that their reforms would lead to economic prosperity. My principal concern, in this paper, has been with Condorcet's policies; with the road not taken, or the *laissez-faire* that was not to be. Malthus's view of Smith has far more influence than Condorcet's on subsequent economic thought. But I would like to suggest, in conclusion, that the effort to reconstruct other prospects — to look back, beyond the political discontinuity of the French Revolution, at the economic ideas of the 1780s — can be enlightening for modern debates.

There is very little support in this early political economy, in the first place, for the view of modern proponents of free market economics that social security is inimical to economic development. Adam Smith's own opinions about the benefits 'for the society' of high wages, and about the civic existence of the poor, lend no credence to a presumption of the enlivening effects of fear; or to the presumption of some of Smith's modern supporters (Mrs Thatcher, for example) that 'there is no such thing as society'. Condorcet and Turgot have at least as good a claim as Malthus to represent the true inheritance of Smith's free market theories. Social justice is not, for these early exponents of *laissez-faire*, 'an ideal that condemns modern commercial society', in the words of a recent report; it is 'an economic necessity', and 'something that society requires because everyone's quality of life is dependent in part on a high degree of social well-being' (Commission on Social Justice, 1994: 19).

The arguments of the 1770s and 1780s in favour of social security are of some interest, secondly, for modern economic policies. Smith thought that the well-being of the poor was both an end in itself and a means to the end of public prosperity; it was 'but equity, besides'. For Condorcet, social well-being was a constituent of the well-being of individuals, and 'the idea that there exist a hundred thousand unhappy people around us is a painful experience just as real as an attack of gout' (Condorcet, 1847–9: V, 361). For David Hume, in his essay 'Of Commerce', such well-being was simply suited to human nature: 'a too great disproportion among the citizens weakens any state. Every person, if possible, ought to enjoy the fruits of his labour, in a full possession of all the necessaries, and many of the conveniences of life. No one can doubt, but such an equality is most suitable to human nature' (Hume, 1987: 265).

Social security was at the same time an important means to economic transformation. Turgot and Condorcet were convinced that some sort of minimum income security was a condition for economic development. They were preoccupied (as was Malthus) by the psychological and institutional conditions for a transition to free markets. Turgot concluded that when the people were so poor as to be subject to periodic crises of their very subsistence, then conditions were unpropitious for enterprise, risk and stable market institutions. People do not feel secure in such a society, or willing to risk the overthrow of old, oppressive institutions. Conditions of social insecurity are unpropitious, too, for public instruction. Enlightenment or education — and the prospect that everyone, including the very poor, will be able to see through interested arguments in favour of government regulation — was for Turgot and Condorcet of critical importance to free market reforms. But the poor could only educate their children if they had some minimum security of income, and people could only remain educated, or instructed, if they had some minimum leisure. 'Enlightenment' was a condition for economic reform, as well as its consequence.

There is some eighteenth century evidence, thirdly, that the provision of minimum security for the poor, and their consequent integration in local or national society, are indeed associated with economic development. Holland and England, the two great commercial empires, were also the two countries most famous (or notorious) for their systems of poor relief. The relation between commercial prosperity and the political condition of the poor was a subject of intense interest at the time, as it was for Malthus and George Ensor. One possibility, as Richard Smith has written, is 'that causation runs from the poor-relief and the minimization of risk and uncertainty that it entailed to economic success, as well as (or instead of) from economic success to poor-relief' (R. Smith, 1986: 206). This was the assumption, at least, of Turgot and of Condorcet. It was the assumption, too, of other contemporary theories of political and economic transformation. 'Through their dependence on chance men become frivolous and idle', as in Naples, Hegel wrote in his account of poor relief: 'in England, even the very poorest believe that they have rights; this is different from what satisfies the poor in other countries' (Hegel, 1821/1967: 277–8). The Prussian reformer Schön attributed his involvement with the legal (and commercial) reform edicts of 1807 to a visit to the English countryside: 'It was through England that I became a statesman. Where the labourer, busy among the cabbages, called out to me in exultation that he had read that my King was about to join the Coalition against France along with England — there you have, in the truest sense of the word, public life' (quoted in Seeley, 1878: I, 376).

The politics of social security, finally, is strikingly different in eighteenth century and in modern debates. All the economists with whom we have been concerned were critical of state regulations. Condorcet's ideal, as late as November 1792, was of a 'virtual non-existence' of the state, based on 'laws and institutions which reduce to the smallest possible quantity the actions of government' (Condorcet, 1847–9: X, 607). But they were even more critical of the existing 'intermediate' institutions of religious, guild, local and parish regulation, whereby the poor were at the mercy of the 'corporation spirit' of the guilds, or of what Smith described as 'the caprice of any churchwarden or overseer' (Smith, 1976a: 154). They were favourable, in principle, to individual charity, but quite unimpressed (as in the French crisis of 1770) by its adequacy as a system of social security. They looked forward to the security of individual rights in a new, more enlightened state.

The characteristic of 'Smith and his disciples', Carl Menger wrote in 1883, is their 'one-sided rationalistic liberalism, the not infrequently rash effort to do away with that which has endured ... the just as rash urge to create something new in the domain of political institutions'. The 'Anglo-French Epoch of Enlightenment', Menger said, can be charged with a 'pragmatism' which 'did not know how to value the significance of "organic" social structures ... and therefore was nowhere concerned to conserve them' (Menger, 1883: 201–2, 207). The early *laissez-faire* economists were not conservatives, on this view. They saw no prospect, in particular, that the existing social institutions of local, religious and corporate charity could constitute the foundations of social equality. There was no perpetual model, that is to say, of optimal institutions for social security, true of all societies, all times, and all configurations of political power. There was rather, for Condorcet, a process of perpetual political consultation, including the consultation, foreseen by Turgot, of people without political power. This was one of the 'pragmatic' ideals of political economy before the French Revolution; it is a reasonable ideal, still, for the politics of social development.

REFERENCES

Académie des Sciences, Arts & Belles Lettres de Châlons-sur-Marne (1779) *Résumé des mémoires sur les moyens de détruire la mendicité en France*. Châlons-sur-Marne: Seneuze.

Baker, Keith Michael (1975) *Condorcet: From Natural Philosophy to Social Mathematics*. Chicago, IL: University of Chicago Press.

Bloch, Camille (1908) *L'assistance et l'état en France à la veille de la Révolution*. Paris: Alphonse Picard.

Burke, Edmund (1894) 'Thoughts and Details on Scarcity', in *The Works of the Right Honourable Edmund Burke*. Boston: Little, Brown.

The Commission on Social Justice (1994) *Social Justice: Strategies for National Renewal*. London: Vintage.

Condorcet (1847–9) *Oeuvres*, edited by M. F. Arago and A. C. O'Connor. Paris: Didot.

Ensor, George (1818) *An Inquiry Concerning the Population of Nations: Containing a Refutation of Mr. Malthus's 'Essay on Population'*. London: Effingham Wilson.

Ferdinand-Dreyfus (1903) *Un philanthrope d'autrefois: La Rochefoucauld-Liancourt, 1747–1827*. Paris: Plon.

Fetter, Frank Whitson (1957) *The Economic Writings of Francis Horner in the Edinburgh Review 1802–6*. London: London School of Economics.

Godwin, William (1820) *Of Population: An Enquiry Concerning the Power of Increase in the Numbers of Mankind, Being an Answer to Mr. Malthus's Essay on that Subject*. London: Longman.

Goethe, J. W. (1962) *Italian Journey* (trans. W. H. Auden and Elizabeth Mayer). London: Collins.

Hazlitt, William (1807) *A Reply to the 'Essay on Population' by the Rev. T. R. Malthus*. London: Longman.

Hegel, G. W. F. (1821/1967) *Hegel's Philosophy of Right* (trans. T. M. Knox). Oxford: Oxford University Press.

Horner, Francis (1806) 'Playfair's Edition of Wealth of Nations', *The Edinburgh Review* VII(XIV) (January): 471.

Hume, David (1987) *Essays Moral, Political, and Literary*, edited by Eugene F. Miller. Indianapolis: Liberty Classics.

Malthus, Thomas Robert (1986) *The Works of Thomas Robert Malthus*, edited by E. A. Wrigley and David Souden. London: William Pickering.

Menger, Carl (1883) *Untersuchungen ü. d. Methode der Socialwissenschaften, u. der Politischen Oekonomie*. Leipzig: Duncker & Humblot.

Mossner, E. C. and I. S. Ross (eds) (1987) *The Correspondence of Adam Smith*. Oxford: Oxford University Press.

Oxenford, John (trans.) (1974) *The Autobiography of Johann Wolfgang von Goethe*. Chicago, IL: University of Chicago Press.

Playfair, William (1805) *The Wealth of Nations (Annotated Eleventh Edition)*. London: T. Cadell and W. Davies.

Robertson, William (1769) 'A View of the Progress of Society in Europe', in *The History of the Reign of the Emperor Charles V*, Vol I. London: Strahan.

Rothschild, Emma (1992a) 'Commerce and the State: Turgot, Condorcet and Smith', *Economic Journal* 102(414): 1197–1210.

Rothschild, Emma (1992b) 'Adam Smith and Conservative Economics', *Economic History Review* XLV(1): 74–96.

Say, Jean-Baptiste (1972) *Traité d'économie politique*. Paris: Calmann-Lévy.

Seeley, J. R. (1878) *Life and Times of Stein, or Germany and Prussia in the Napoleonic Age*. Cambridge: Cambridge University Press.

Simonde de Sismondi, J.-C.-L. (1819) *Nouveaux principes d'économie politique, ou de la richesse dans ses rapports avec la population*. Paris: Delaunay.

Skinner, Andrew S. (1993) 'Adam Smith and the Role of the State: Education as a Public Service' (mimeo). University of Glasgow.

Smith, Adam (1976a) *An Inquiry into the Nature and Causes of the Wealth of Nations*, edited by R. H. Campbell and A. S. Skinner. Oxford: Clarendon Press, 1976.

Smith, Adam (1976b) *The Theory of Moral Sentiments*, edited by D. D. Raphael and A. L. Macfie. Oxford: Clarendon Press, 1976.

Smith, R. M. (1986) 'Transfer Incomes, Risk and Security: The Roles of the Family and the Collectivity in Recent Theories of Fertility Change', in David Coleman and Roger Schofield (eds) *The State of Population Theory: Forward from Malthus*, pp. 188–211. Oxford: Basil Blackwell.

Southey, Robert (1832) *Essays, Moral and Political*. London: John Murray.

de Stendhal (1990) *Le rouge et le noir*. Paris: Pocket.

Turgot, A. R. J. (1913–23) *Oeuvres de Turgot et Documents le Concernant*, edited by Gustave Schelle. Paris: Alcan.

Wrigley, E. A. (1987) *People, Cities and Wealth: The Transformation of Traditional Society*. Oxford: Basil Blackwell.

Traditional Co-operatives in Modern Japan: Rethinking Alternatives to Cosmopolitanism and Nativism

Tetsuo Najita

We live at a time when our social and political environments are being tested by ambiguous and contradictory pulls. New technologies generate a consumerism that extends beyond national boundaries. The globe, we are constantly being reminded through the media, continues to shrink into a new kind of oneness. Yet even as the world undergoes change at a bewildering pace, with each new 'upgrade' there is a counter tendency to retrench, to rediscover firm boundaries closer to home and to national and ethnic place, to insist on distinctiveness and difference in defiance of globalism or even of regionalism. Old boundaries resurface to be contested, so that they can serve as stable and controllable lines. These claims are invariably tied in complex and subtle ways to the yearning to be creative and to be distinctive in the face of technological surges toward homogenization.

The more relentless and aggressive these surges, the stronger the social desire to identify clear limits nearby and to insist on an unambiguous identity that flows in an essential mainstream from the past. Unfortunately this resistance to the mechanical erasure of boundaries and affirmation of the socially familiar can often deteriorate into erratic and inflammatory rhetoric and the rekindling of old flames of distrust and passionate hostility. As technology expands, there seems to be no stable place for basic human values in modern consciousness.

When we discuss social development, our intellectual groundings can pull us toward a kind of institutional- and knowledge-culture which might be described as 'cosmopolitan'. This approach, with its generalizing tendency, often dehistoricizes social processes and refuses them an authentic place. Our theoretical range is thus curtailed by epistemological boundaries, 'reduced formulas', in the wording of Clifford Geertz (1973), that permit little or no understanding of the problems of knowledge and commitment and contract and practice in other societies. The 'other' is viewed in a limited way, 'thematized' (drawing on the wording of Wlad Godzich) 'as a threat to be reduced, as a potential same-to-be, yet-not-same' (de Certeau, 1986: xiii).

This cosmopolitan 'gaze' is often internalized by those under scrutiny in other countries, generating an urge on their part to explain to outside audiences 'who they really are' in some comprehensive culturalist or historicist terms. At the same time, their explanation may strike an elitist stance that criticizes the rest of their own society as intellectually lacking. When this occurs, the specific concerns which ordinary citizens of that country may have about excessive growth, ecological damage and discrimination of one kind or another are glossed over, as are the solutions to these issues which may be worked out in rather unspectacular ways, as people question the relative merits of modern growth.

In contrast, the counter-gaze to cosmopolitanism, which is nativism, insists that a society should view itself in terms of a stable and unchanging identity, rather than as an entity to be described as 'yet-not-same'. By disavowing change, however, nativism also denies a country its history; and thus it, too, confounds our understanding of social development. In its claim to a distinct identity, nativism may refer to an autonomous 'language' as being the essential core of a society and thus reference cultural identity to poetics and oral and mythic traditions. Or it may reduce cultural essence to an absolute political identity and, in turn, find a congenial bed-fellow in the Western concept of national sovereignty, which in its most extreme position recognizes no formal constraints except perhaps a superior coercive power. There is also the more subtle and complex variant of nativism, when a self-conscious aesthetic preference is seen as the only possible avenue for a country to be creative in a world driven toward sameness.

One of the more trenchant comments in support of the latter position was made by the writer Tanizaki Jun'ichiro in his essay *In Praise of Shadows*, written in the early 1930s. To choose the dimness and shadows of the Japanese toilet over the modern, Western, well-lit room was a matter of aesthetic preference, he wrote, a way to begin one's day without the interference of cosmopolitan reason. While technology was unavoidable, inevitable as is the movement of time, one ought still to preserve an aesthetic space of one's choosing, a matter of one's preference, with utter indifference as to what anyone might think — an inner space, he would say, 'where we can turn off the electric lights and see what it is like without them' (Tanizaki, 1977: 48).

*****

Both the hierarchical cosmopolitan view of societies, considered to be in relative states of unreadiness, and the nativist insistence (with its nuanced gradations) on distinctiveness persist as seductive sense-making devices. They are perhaps unavoidable. Yet it seems to me that if we seek a historical perspective on social development, we need to acknowledge these epistemologies and then put them aside. Borrowing Tanizaki's phrasing, we need to turn off the electric lights and see social development without them — so that we can observe the dimly lit social spaces in which human beings act out a productive history.

To understand the social development of a country, it is important to focus upon the thought and practices of ordinary people in the everyday world of ethics, contract and commerce. In other words, we need to view social spaces in terms of what is being created by the people of that society, without prejudging this field of action as modern or traditional or Eastern or Northern. This is a history that is relatively devoid of dramatic and extreme events, and of egregious biographies, since it is largely (with perhaps an exception or two) about nameless people. Although it is a history that has no readily apparent narrative, it is grounded in discursive practices that manage to survive the impact of events and the interventions of modern apparatuses and ideologies. Nevertheless these practices do not persist intact. They are scattered about in fragments which are often not readily noticeable, except as traces of unadvertised practices; and they are sometimes even self-consciously hidden from view. It is in this amorphous arena of everyday thought and practice, where determinations are made as to why and how things are to be done in order to live through a personal history, that a perspective on social development can be shaped.

*****

For the past several years I have been tracing the history of contract co-operatives in Japan from the pre-industrial eighteenth century into the modern era. They are not part of the narrative of modernization. They are not put forth as evidence of cosmopolitan pride. They are not part of the history of state formation. They do not reveal economic individualism according to the market model of society. Nor are they referenced to the native soul, as a confirmation of differential identity.

The contract co-operative is, broadly speaking, Asian; in this regard it is not uniquely Japanese. Its origins go back to informal reading and prayer groups, or co-operatives formed to sustain the elaborate Buddhist temple establishments in Asia in the seventh and eighth centuries. It was in the eighteenth century, however, that co-operatives became broadly secularized in Japan, as commoners acted jointly in mutual commerce, venture and insurance. The co-operatives thus became part of Japanese social history.[1]

---

1. For more on these co-operatives, see for example Mori Shizuro (1978–81) and Mori Kahei (1982).

In the late 1940s, this development was referred to as 'co-operative' democracy or sometimes simply as 'social' democracy, but in the 1960s and 1970s the phraseology dropped out of sight as liberalism came to prevail under conditions of high economic growth. Still, the co-operatives remain everywhere in Japan, beneath the internationally oriented economic organizations, and their history is one we might consider in our discussion of social development.

Two prominent and mutually entwined forces — one philosophical, the other practical — were behind these eighteenth-century contract co-operatives. Philosophically, the formation of co-operatives was supported by the spread of a theory of knowledge which held that the natural order was not based upon a dualism of spirit and matter, physical and metaphysical, but upon a unified, monistic and infinite material energy, a ceaseless process of motion with neither beginning nor end (as expounded by Ito Jinsai, 1627–1705). What this seemingly heady philosophy did was to confirm certain powerful ethical preferences: since natural energy could not be deemed evil, there were no independent forces of good and evil in human society — only goodness or the relative propensity in humans to practise it. Similarly, there was no constant of male and female dualism but only human life, the gender distinction always being reduced to a life principle; and there was no distinction between life and death which would imply that life is impermanent, while death is transcendent and eternal. Death was contained within the monistic, life-only process.

The ramifications of this philosophic monism reached into practical spheres. It confirmed that the energy within each being was a fundamental 'gift' (*on*) of the natural order, and reverence for that gift required individuals to nourish life through work and ethical practice in the everyday world. Commoners in their various fields of work came to be empowered with a sense of moral integrity. In cities such as Kyoto and Osaka, merchants used this theory of knowledge and practice to establish schools and academies. At the Osaka Merchant Academy, for example, the human virtue of acquiring knowledge of all kinds (moral, political and economic) was taught to be an intrinsic capacity in everyone, not a special prerogative of the aristocracy. This perspective reinforced a political theory accepted along the entire intellectual spectrum: that governments existed for one purpose only, to nourish human virtue, or 'to order and save the people', as this was phrased at the time.[2]

In the countryside, an agrarian communalism proposed that all dualisms, such as those confirming status and gender distinctions, should be abolished and replaced by direct agricultural work within the vast interconnectedness of the natural order (Ando Shoeki, 1703–62). Profound epistemological

---

2. The root ideography of the modern term for 'economics' is to order and save, or *keizai*. The ethic of saving others to which these ideographs refer had nothing to do, originally, with the modern economic concept of profitability.

discussions accompanied this critique of philosophical dualism. Conventional language itself was criticized as a source of deception, since received language was invariably laden with customary beliefs and habits of mind that severely limited an understanding of nature. Perhaps, it was argued (for instance by Miura Baien, 1723–89) the basic categories of knowledge embedded in conventional language needed to be drastically altered.

Armed with these views, intellectuals dared to challenge the official policy of seclusion and to explore Dutch medicine. Dutch studies in Japan were driven by the desire to find firmer groundings, especially in the science of human anatomy, to achieve the ethical goal of saving human lives. Thus began the tradition from the 1750s onward of studying the Dutch language and translating Western works on science. Significantly, to respond adequately to the gift of life implied nourishing it through systematic agricultural work, since the unfolding of the natural order was not arbitrary but somehow principled and accurate. This became the anchor for the development of Tokugawa agronomy, or scientific farming.[3] Agricultural work was granted an ethical dignity inferior to no other form of human endeavour, for the natural, material world of the unfolding of things was one with the seemingly pure and spiritual realm of the heavenly and timeless (Kaibara Ekken, 1639–1714).

Based on this philosophy of reverence for the immediate and concrete life as a gift of the natural order, as well as on the obligation of human beings to nourish that gift, the origins of contract in Japan were ethical principles. They were located in a promise, oath or agreement by each person in a township or village community to save any another under conditions of extreme emergency, so that when there was a famine, no one would be neglected or allowed to die.[4] The oath may be linked to the political concept mentioned above — that governments existed to save the people — except that here, commoners seized the initiative to save themselves from below.

Rather than waiting for nourishment to filter down from the feudal régimes, commoners who founded contract co-operatives pledged to save each other, because the promises that gave politics ideological legitimacy went regularly unfulfilled. The contract co-operative, in short, was grounded in a disbelief in politics among commoners, a disbelief which should not be confused with mere indifference and certainly not with obsequiousness. All villages without exception maintained a contract co-operative to provide security in times of fire, flood and famine.

The following extracts come from a village contract of the mid 1750s, written in a language accessible to the peasantry itself:

Those who eat beyond being full, and who overly bundle themselves in warm garments and simply look with indifference as others die before their very eyes, will incur the wrath of

3. For more on this, see Tetsuo Najita (1987, 1994).
4. See also Charles Fried (1981).

heaven and, in the social world around them, will have divorced themselves from the ways of human ethics ...

Think of the misfortunes that happen to others around you as though they were your own. In the everyday course of events, live with care and attention, interact with others in the village as do the fish who swim in water. Each will then be said to receive the blessings of heaven.

[It is agreed therefore that] those who are able to read will explain in detail the above principles [of saving each other] to their spouses and children, and to all others in their households ...

These were thus moral contracts. They did not provide legal mediation in economic exchanges. Indeed, the history of contract co-operatives took place almost entirely beyond the public order, beyond the reaches of political direction or juridical regulation. Since the aristocracy did not live in the villages but in castle towns, commoners wrote these agreements for themselves without leadership or direction from above. They are legitimated, as in a Kantian sense perhaps, by the moral promise not to treat others as mere objects, but as fellow human beings worthy of being saved. Many of these contract co-operatives referred to themselves, quite appropriately, with the Buddhist ideographic compound meaning 'unlimited compassion' (*jihi mujin*).

This very virtuous content conveys the impression of vagueness and emotionality, and hence of being out of step with modern reason. Invariably overlooked, however, is the fact that the contracts rested on an epistemological commitment to accuracy. As little as possible was left to whim and chance. Risk was to be minimized and savings were to be made systematic. Not meeting one's responsibilities in concrete and quantifiable terms without just cause was indication of moral failure, and hence of a much diminished sense of one's humanity. Abiding by the terms of accuracy in order to save others determined one's human dignity and selfhood. Indeed, the commitment to accuracy might be seen as preceding trust; or to reverse this statement, without accuracy there could not be trust. The ideograph for 'righteousness' and 'justice' was often used to mean mathematical accuracy.

It is not surprising, then, that village contracts would contain within them an interest-bearing loan and trust fund. Terms were set for collecting cash, keeping detailed records, requiring security deposits, prohibiting the use of funds for personal profit, and maintaining the trust fund into the indefinite future, as long as the community survived. I shall again quote from the previous village contract:

Those whose contributions accumulate to a sum in excess of 10 [units of cash] ... or 10 [units of grain] ... will be designated benefactors; their names will be duly recorded in the benefactors' book of accounts; and their funds will then extend indefinitely into the future to aid all those who encounter extreme and unexpected disasters.

As these terms indicate, the contract was for the long haul, from one generation into the indefinite future, and it was linked at its core to an interest-bearing fund. Contract co-operatives therefore persisted into the modern era and were especially important during times of severe epidemics such as cholera and measles. They were also adaptive and expansionist in their activities.

The contract co-operative served, for example, as the organizing principle for village reconstruction movements in the early 1800s — known most prominently as the Hotoku movement. Villages were linked by reliance on identical contracts and by agreement on the ethical purpose underlying them. The objective of these contracts was not so much to react to famine as to avoid it altogether through systematic improvement of agricultural production, savings and budgeting, and the establishment of an interest-bearing credit and loan fund. The fund, which grew mainly through regular savings deposited by members, extended loans on an 'interest-free' basis. An assumption accompanied these loans to the effect that repayments would include an additional year of compensation beyond the loan period. The language 'interest-free' loans, however, was protected with a fierce sense of pride to safeguard the ethical principle that the fund was dedicated to saving others without expectation of repayment. Accordingly, a portion of savings was to be set aside for philanthropic purposes, or 'returning the gift of life to society'.

In the early modern industrial era, around 1910, this village reconstruction movement would not accept the state's guidelines that credit and loan facilities operate on the modern, rational principle of profitability. Instead the counter argument was made that the ethical principle of saving to help others must remain central to and inseparable from economics. Granting interest-free loans was a prerogative of the movement itself, not a matter to be determined by the state. Today, two stone columns at the entrance to the headquarters of the Hotoku movement, marked 'moral pillar' and 'economic pillar', are clear reminders that it is only modern society that makes a distinction between the two.

Using the same basic concepts, contract co-operatives expanded in scope and became venture oriented. Unlike open-ended insurance contracts, venture-oriented contracts were for limited time periods — say ten years. Funds could be used on a rotational basis for any purposes, whether for family or personal needs or for investment. In these limited contract co-operatives the ethic of reciprocal aid and reliance was retained, so that within the contracting group each participant was at once creditor and debtor to each of the other members. Those who had drawn from the pool of funds became debtors to those who had not yet drawn, and as such they continued to deposit into the pool until all obligations had been fulfilled. In the absence of banks, these contract co-operatives served as credit and loan facilities in which individuals borrowed from each other.

In the early industrial era of the 1880s and 1890s, contract co-operatives were articulated as companies. Hundreds of small, limited-contract co-operatives were pooled into a single entrepreneurial entity; through a variety of twists and turns, they formed the bases of the mutual trust banks and credit associations that are today scattered throughout Japan's urban neighbourhoods and small towns. The idea and practice of mutual insurance (*sogo fujo*) continue to be basic ethical components in these popular institutions. Around 1910, the Ministry of Finance decided to do a survey to see exactly how many of these contract credit co-operatives existed in Japan. The final count, published in 1914, was 333,634. Almost all of these functioned outside of the legal or public order, without state direction or regulation, thus retaining their commoner origins.

* * * * *

As we discuss social development in a global context, it is perhaps important to note the strong impression that the co-operative movements of Northern Europe made on students from Japan studying there in the 1870s and 1880s. What struck them about local and regional communities in that part of Europe was not the prevalence of economic liberalism but the vitality of co-operatives, reminding them very much of mutual aid and venture activities back home. The ideas identified with Friedrich Wilhelm Raiffeisen (1818–88) and Franz Hermann Schultze-Dielitzch (1808–83) were to become part of the discourse on co-operatives in Japan. Some of these students emphasized the importance of legislation to make indigenous co-operatives legal within the constitutional order; others focused on co-operatives as commoner banks; still others were impressed by the theme of the producer-first, rather than management-first, principle of economics; and many found encouragement in the ethics of Christian humanitarianism that dovetailed with their own understanding of compassion and mutual assistance.

One of these students was Kurosawa Torizo, whose life (1885–1982) spanned almost one hundred years of Japan's modern history. Born in the 1880s, he was educated in the natural order and ethical action philosophies that were received from the old régime. He went on to study mathematics in Tokyo, but then (around 1900) joined the first protest movement against ecological damage caused by copper poisoning in the rice fields. His mentor and leader of the protest movement (Tanaka Shozo, 1841–1913) hired a lawyer to have him released from prison and sent him north, to the distant area of Hokkaido, where friends could help him start a new life. While his mentor met his death in the poisoned rice fields, Kurosawa Torizo distanced himself from his studies of mathematics and began to learn about dairy farming. His studies included an extended trip to northern Europe, and in particular to Denmark, where he focused his attention on the co-operative movement. Throughout his life he consistently held Denmark in high esteem, celebrating it in some of his poetry, for example, as a modern society that had transformed itself through its agricultural industries.

In the early 1920s, Kurosawa Torizo founded a dairy co-operative in Hokkaido, based on the producer-first principle rather than on the profitability of individual ownership, and incorporated the ethical principles of mutual aid and saving strategies practised in the contract co-operatives of the old régime. To the very last of his many lectures, in the early 1980s, he gave credit for his work to both the ecology and co-operative movements in which he had participated as a youth. I might add that the company he founded remains a thriving entity today.[5]

I should like to emphasize one further point. The ideas and practices of co-operatives were viewed, especially in the 1910s and 1920s, as essential elements in modern democracy. Even those modelled entirely on pre-modern contract co-operatives were not thought to be 'traditional', but 'democratic'. More than individual advancement based on economic achievement, democracy was believed to include the horizontal principle of equality and of mutual aid and sharing. I was very much struck by this connection recently when I interviewed a woman in her mid-seventies who had founded a co-operative in Tokyo in 1946, immediately after the end of the Pacific war. Under conditions of enormous hardship in which many youngsters (including two of her own children) died of malnutrition and lack of medical supplies, it was, she noted, the most natural and democratic thing to do. She was simply practising the democratic ideals of equality and mutual aid as they had been taught to her as a student at Nara Higher School for Women, Nara Joshiko.

Scattered about on the traditional matted floor of her house were computer printouts on nutrition and world trends in food prices. She noted that economists conducted regular seminars for her organization because the co-operative was not interested in charity, but in improving the quality of human life; and they had no intention of losing out to profit oriented economics. She went on to say, in parting, that the success of her co-operative rested on the fact that at its core was the contract co-operative of 'unlimited compassion', traceable to the eighteenth century.

* * * * *

In closing, I would like to make it clear that I am not trying to suggest that we should reconstitute the history of Japan's old feudal régime and the community virtues of the time: we need only remind ourselves that the 'contract' among commoners was a counter-strategy to famine, a condition that I certainly do not wish to romanticize. Nor do I want to suggest that mutual trust banks and co-operative credit associations occupy a substantial part of Japan's industrial economy. That they served as a crucial credit infrastructure for ordinary citizens under capital poor conditions, however, is beyond doubt.

My more important point is that this credit and insurance structure was not provided by the state; it adapted and expanded beyond the public order,

---

5. For more on Kurosawa Torizo, see the biography by Aoki Hisashi (1961).

outside legal regulation. It is thus a history that cautions us about depending excessively upon the overall competence of modern states in providing health and economic care for the people. From this perspective, the contract co-operative is not simply a hangover from a feudal or 'Asian' tradition, but an institution within social history which reflects a society's aspiration to provide for the basic human needs of its members. Social development is about the creation of social spaces through such efforts, and our historical narratives must take these actions into account. The categories of 'modern' and 'traditional' continue to limit how we view and assess contemporary experience — thus the 'predicament of modernity'. When 'modern' is cosmopolitan reason and 'traditional' is native identity, especially in its most militant manifestations, we lose sight of a vast area of social practice in which the basic starting point is maintaining life.

The conception that nature is one — that each individual living thing (*hitori*) is in motion, interconnected in an infinite process encompassing life and death — is worthy of consideration in our times. As stated, this monistic philosophy denies dualistic distinctions of high and low, male and female, heaven and earth, life and death, and, finally, admits only 'motion' and 'rest' in a ceaseless process without beginning or end. It is a monistic philosophy that has been largely forgotten in modern Japan. This is unfortunate since there is much in the naturalist philosophy underlying agronomy and contract insurance co-operatives that speaks to some of that country's central issues and ought to be retrieved. Perhaps this will happen. As citizens' movements reflect deep concern about the excesses of technological efficiency, and the ecological damages that ensue, questions are being raised about the relevance of historical resources from an earlier time — among these, the eighteenth-century theory of the natural order, the ethic of reverence for life and the imperative of saving one another.

REFERENCES

Aoki Hisashi (1961) *Kurosawa Torizo* (biography of Kurosawa). Sapporo: Torizo den Kankokai.
de Certeau, Michel (1986) *Heterologies. Discourse on the Other* (with foreword by Wlad Godzich). Minneapolis, MN: University of Minnesota Press.
Fried, Charles (1981) *Contract as Promise.* Cambridge, MA: Harvard University Press.
Geertz, Clifford (1973) *The Interpretation of Cultures.* New York: Basic Books.
Mori Shizuro (1978–81) *Shomin Kin'yu shisoshi taikei (An outline intellectual history of commoner credit co-operatives),* 3 vols. Tokyo: Nihon keizai hyoron sha.
Mori Kahei (1982) *Mujin Kin'yu shi ron, Chosakushu (A study of eighteenth-century credit co-operatives of unlimited compassion),* Vol. 2. Tokyo: Hosei daigaku shuppan.
Tetsuo Najita (1987) *Visions of Virtue in Tokugawa Japan.* Chicago, IL: University of Chicago Press.
Tetsuo Najita (ed.) (1994) *Tokugawa Reader.* Chicago, IL: University of Chicago, Center for East Asian Studies.
Tanizaki Jun'ichiro (1977) *In Praise of Shadows (In'ei raisan,* trans. by Thomas J. Harper and Edward G. Seidensticker). New Haven, CT: Leete's Island Books, Inc.

Affluence, Poverty and the Idea of a Post-Scarcity Society

Anthony Giddens

The starting point of this discussion is a world that has taken us by surprise. By 'us' I mean not only intellectuals and practical policy-makers, but the ordinary individual too. In the West, at least, we are all the legatees of certain strands of Enlightenment thought. The Enlightenment was a complex affair. Various different perspectives of thought were bound up with it and the works of the leading Enlightenment philosophers were often complex and subtle. Yet in general the philosophers of Enlightenment set themselves against tradition, against prejudice, and against obscurantism. For them the rise of science, both natural and social, would disclose the reality of things.

Understanding was always itself understood as an unfinished and partial affair — the expansion of knowledge is at the same time an awareness of ignorance, of everything that is not and perhaps will not be known. Nevertheless, knowledge was presumed to be cumulative and presumed also to yield a progressive mastery of the surrounding world. The more we are able to understand ourselves, our own history, and the domain of nature, the more we will be able to master them for our own purposes and in our own interests. The underlying theorem, stripped bare, was extremely plausible. The progress of well-founded knowledge is more or less the same as the progressive expansion of human dominion.

Marx brought this view its clearest expression, integrating it with an interpretation of the overall thrust of history itself. In Marx's celebrated aphorism, 'human beings only set themselves such problems as they can resolve'. Understanding our history is the very means of shaping our destiny in the future. Even those thinkers who took a much less optimistic view than Marx of the likely future for humanity accepted the theorem of increasing human control of our life circumstances. Consider, for example, the writings of Max Weber. Weber certainly did not see history as leading to human emancipation in the manner envisaged by Marx. For Weber, the likely future was one of 'uncontrolled bureaucratic domination' — we are all destined to live in a 'steel-hard cage' of rationality, expressing the combined influence of bureaucratic organization and machine technology. We are all due to be tiny cogs in a vast and well-oiled system of rational human power.

Each of these visions of the imminent future attracted many adherents. Marxism, of course, shaped the very form of human society for many. Others, perhaps critical of Marxist thought, recoiled before the sombre vision offered by Weber, Kafka and many others. Marxism, as we all know now, has lost most of its potency as a theoretical perspective on history and change. But Weber's more sombre vision has also lost its hold over us. It does not correspond to the world in which, at the end of the twentieth century, we in fact find ourselves. We do not live in a world which feels increasingly under human control but, rather to the contrary, one which seems to run out of control — in the words of Edmund Leach (1968), a 'runaway world'. Moreover, this sensation of living in a world spinning out of our control can no longer be said to be simply the result of lack of accumulated knowledge. Instead, its erratic runaway character is somehow bound up with the very accumulation of that knowledge. The uncertainties which we face do not result, as the thinkers of Enlightenment tended to believe, from our ignorance. They come in some substantial part from our own interventions into history and into the surrounding physical world.

It is unlikely that the world in which we live today is actually more uncertain than that of previous generations; it would be hard, in any case, to see how such a claim could be validated. It is the sources of uncertainty which have changed. We live increasingly in a social and material universe of what I shall call 'manufactured uncertainty'. Manufactured uncertainty, or manufactured risk, comes from human involvement in trying to change the course of history or alter the contours of nature. We can separate manufactured risk from 'external risk'. External risk refers to sources of uncertainty which come either from unmastered nature or from 'unmastered history' — that is, history as lived by taken-for-granted traditions, customs and practices.

The debate about global warming — which is a debate about 'nature that is no longer nature' — offers one among many examples of the advent of manufactured uncertainty. The majority of scientific specialists believe that global warming is occurring, even if all forecasts of its likely consequences are imponderable. Some scientists, however, believe that the whole idea of global warming is a myth, while there is a minority view that what is taking place is actually the reverse — a long-term process of global cooling. The uncertainties which surround the global warming hypothesis do not derive from 'unmastered nature', but precisely from human intervention into nature — from the 'end of nature'. Since we cannot be wholly sure whether or not global warming is occurring, it is probably best on a policy level to proceed in an 'as if' manner. As some of the consequences of global warming could be calamitous, it is sensible for nations and the larger world community to take precautionary measures.

Manufactured uncertainty is by no means limited to 'nature which is no longer nature'. It invades most areas of social life too, from local and even personal contexts of action right up to those affecting global institutions. Take as an example the decision on the part of someone living in a Western society to get married. Fifty years ago, someone who decided to marry knew what he or she was doing: marriage was a relatively fixed division of labour involving a specified status for each partner. Now no one quite knows any longer what marriage actually is, save that it is a 'relationship', entered into against the backdrop of profound changes affecting gender relations, the family, sexuality and the emotions.

What explains the increasing dominance of manufactured over external risk? Obviously the origins of this transition are bound up with the advent of modernity as a whole. However, a series of very basic changes sweeping through the world over the past several decades have intensified this transformation of the conditions of uncertainty and risk. Three particular sets of changes are having a major impact throughout the industrialized countries and are, in some degree, affecting most societies across the globe.

The first set of changes concerns the effects of globalization. The word globalization appears almost everywhere these days, but thus far has not been well conceptualized. As I would understand it here, globalization does not simply refer to the intensifying of world economic competition. Globalization implies a complicated set of processes operating in several arenas besides the economic. If one wanted to take a technological fix upon the intensifying of globalization in recent years, it would be the point at which a global satellite communication system was first established. From that point onwards instantaneous communication became possible from any part of the globe to any other. The advent of instantaneous global communication both altered the nature of local experience and served to establish novel institutions. The creation of twenty-four hour money markets, for instance, a phenomenon that has an impact upon almost all the world's population, became possible only because of the immediacy of satellite communication.

Globalization is not just an 'out there' phenomenon. It refers not only to the emergence of large-scale world systems, but to transformations in the very texture of everyday life. It is an 'in here' phenomenon, affecting even intimacies of personal identity. To live in a world where the image of Nelson Mandela is more familiar than the face of one's next door neighbour is to move in quite different contexts of social action from those that prevailed previously. Globalization invades local contexts of action but does not destroy them; on the contrary, new forms of local cultural autonomy, the demand for local cultural identity and self-expression, are causally bound up with globalizing processes.

The second major source of social change over recent years is detraditionalization. Here again we can distinguish longer processes of transformation from the more intensified changes happening over the past few decades. Modernity, of course, always set itself against tradition — this was one of the very origins of the Enlightenment. Yet during the lengthy period of what Ulrich Beck has called 'simple modernization', modernity and tradition existed in a sort of symbiosis (see Beck et al., 1994). Science itself became a kind of tradition — an established authority to which one turned when seeking the answer to puzzles or problems. This symbiosis of modernity and tradition marks the phase of 'simple modernization' — roughly speaking, the first century and a half or so of industrialization and modernity.

In the phase of 'reflexive modernization', which has accelerated over the past several decades, the status of tradition becomes altered. Detraditionalization does not mean an end to tradition. Rather, traditions in many circumstances become reinvigorated and actively defended. This is the very origin of fundamentalism, a phenomenon which does not have a long history. Fundamentalism can be defined as tradition defended in the traditional way — against the backdrop, however, of a globalizing cosmopolitan world which increasingly asks for reasons. The 'reason' of tradition differs from that of discourse. Traditions, of course, can be defended discursively; but the whole point of tradition is that it contains a 'performative notion' of truth, a ritual notion of truth. Truth is exemplified in the performance of the traditional practices and symbols. It is not surprising, therefore, that we should see so many clashes and fracturings today across the world as embattled tradition clashes with much more open life-style choice.

Detraditionalization is closely linked to the 'end of nature' and indeed the two intertwine very often. 'Nature' disappears in the sense that few aspects of the surrounding material world — and of the body — remain uninfluenced by human intervention. Tradition and nature, as it were, used to be 'landscapes' of human activity, carrying with them a certain fixity of life-style practices. As tradition and nature dissolve, a whole host of new decisions must be taken (by somebody) in areas which were not 'decisionable' before.

Consider, for example, the field of human reproduction. A variety of aspects of reproduction which were previously 'given' — not open to being influenced by human decision-making — now are in principle or in practice malleable. It is possible to have a child without any kind of sexual contact with another adult at all; the sex of a child can become a matter of choice; contraception becomes highly effective, so that the decision to have a child becomes something quite different from when childbearing was more of a 'natural' process. The 'end of nature' in the domain of reproduction, however, integrates closely with the social changes brought about by detraditionalization. Thus central to the lowered birth rate in the developed societies today is the series of changes which have promoted the autonomy of women and therefore altered the traditionally-given relations between the sexes.

The third set of changes sweeping through the world are those associated with the expansion of social reflexivity. This is again not confined to the Western or developed societies, but is bound up with the globalization of communication. 'Reflexivity' does not mean self-consciousness. It refers precisely to the condition of living in a detraditionalized social order. In such an order everyone must confront, and deal with, multiple sources of information and knowledge, including fragmented and contested knowledge claims. Everyone in some sense must reflect upon the conditions of her or his life, as a means of living a life at all. Consider the example mentioned above — the decision to get married. That decision is taken amid a welter of information about 'relationships', 'commitment', the changing nature of sexuality, of gender relations and of the very institution of marriage itself. Such information or knowledge is not simply a 'background' against which the decision to marry is taken: as noted earlier, it enters constitutively into the environment of action which it describes.

Living in a highly charged reflexive social environment brings many new rewards and forms of increasing autonomy; at the same time, it also brings new problems and anxieties. Consider, as an illustration of this, eating disorders and anorexia. As a widespread phenomenon, eating disorders in Western countries are relatively recent, dating only from the past thirty or so years. They are pathologies of a society where everyone is 'on a diet': that is, a diversity of foodstuffs is available, to those who can afford them, at any time of the day, month or year. Diet is no longer given by 'nature' — by the local seasons and by the availability of local produce. In such circumstances individuals have to decide what to eat — in some sense select a diet — in relation to how they want to be. Diet becomes intrinsically bound up with the cultivation of the body: for some people, particularly young women, social pressures to do with bodily appearance can assume a pathological and compulsive form.

When we decide what to eat, and therefore how to be, we know that we are taking decisions relevant to present and future health. A person might resolutely stick to a traditional diet, continue to smoke and so forth, in the face of widely disseminated medical knowledge which indicates these habits to be harmful. Yet he or she cannot do so without being aware of such knowledge claims. Ignoring them is in effect a decision.

In a globalizing world, marked by the swathes of social change just described, pre-established institutions start to come under strain. This is true of areas of social life ranging from personal and intimate social ties right through to large-scale global orders. In politics, to take one illustration, the voting population now lives in the same discursive arena as their political leaders. In such a circumstance, political legitimacy starts to come under strain. Deference tends to disintegrate, and political activities and procedures which were once acceptable start to be placed widely in question. It is not just happenstance that corruption cases have come to the fore in political life in many countries across the world. Corruption was there previously, although it might not have been treated as such; but in the new conditions of social visibility in which political life operates today, what was once accepted becomes generally seen as illegitimate (although the reverse can also on occasion be true).

Rather than developing the political example, I shall concentrate here upon the question of the welfare state and welfare institutions. Most students of social policy agree that the Western welfare state is in a situation of crisis. That crisis is ordinarily understood in fiscal terms — as part of a 'can't pay, won't pay' mentality on the part of the middle classes. In the more affluent sectors of society, in other words, people increasingly refuse to accept the levels of taxation required to support others less fortunate than themselves. Sometimes the fiscal crisis of the welfare state is described, as in Galbraith's phrase, as a 'culture of contentment': many middle class people have achieved a comfortable way of life and become protective about it. Others see the situation more as one of anxiety and insecurity; the middle class is no longer exempt from worries which used to concern mainly those in the lower strata of the social order.

This is not to say that the thesis of the fiscal crisis of the welfare state, in either of these competing versions, is a wholly mistaken one. It is not. However, one can also look at the problems facing the welfare state in a different way. The crisis of the welfare state, it can be suggested, is in some large part a crisis of risk management. The welfare state originated as a 'security state' and was actually called such in some countries. It was the socialized, public counterpart to private insurance. Now the involvement of modernity with insurance makes an interesting and informative story. Modern civilization on the whole looks towards the future rather than the past, seeking to 'colonize the future'; the future is a 'territory' to be 'occupied'. It is not surprising, therefore, that early industrial enterprise was closely bound up with the emergence of the notion of insurance. What is insurance? It is a means of organizing future time. Insurance is a means of protecting against the hazards which might in the future befall individuals or groups in different contexts.

The welfare state was an insurance system which was developed in terms of coping with external risk. Certain things could befall the individual: he or she could become ill or disabled, be divorced or become unemployed. The welfare state would step in to protect those who fell foul of such contingencies. In an era coming to be dominated by manufactured uncertainty, by contrast, welfare institutions based on external risk start to break down. The changing circumstances of divorce serve as an illustration: half a century or so ago, in most Western countries, only a minority of people got divorced, and most of these were cases of men leaving women, because legal and economic circumstances made it difficult for women to extricate themselves from marriage. Where only few divorced, divorce could be treated like a 'hazard of nature' — it might happen to you if you were very unlucky. Where it did take place, divorce happened against the backdrop of gender and family relations which were quite clearly defined and fixed. Today, not only are divorce rates very high compared to what they were; the large proportion of marriages in Western countries are actively broken up by women. In such a situation, reflecting so many other changes in personal and economic life, treating divorce as a 'hazard of nature' makes no sense. Divorce (and remarriage) become part of a much more active series of engagements with life problems. Welfare systems cannot simply step in to pick up the pieces; they have to be redirected and reorganized in such a way as to promote responsible decision-making.

Something parallel applies in the case of health and illness. The medical health care systems of the welfare state were based upon the assumption that falling ill was something which simply happened to people in certain circumstances. In a world of much more actively organized life-styles, where the body is no longer so much of a 'given', this assumption no longer holds. One's state of health tends to be strongly influenced by the life-style decisions which one takes, and by alterable states of the surrounding environment. Health care systems come under strain not simply because of the escalating costs of standard medical treatments, but because they still depend too much upon the presumption of illness as external risk.

In recent times critiques of the welfare state have come mostly from the neo-liberal Right. Neo-liberals see welfare institutions as promoting dependencies rather than encouraging more responsible life-style practices. The impulse of neo-liberalism has been to cut back upon welfare expenditure and to seek to turn welfare systems into markets wherever possible. In an oblique and negative sort of way, the neo-liberals have had a better grasp of the inadequacies of the welfare state in current social conditions than have most of its defenders. But the relevance of their critiques has been undermined by their fascination with markets. In place of the neo-liberal attack upon welfare institutions, we should seek to provide what could be described as a positive critique of the welfare state, rather than a primarily negative one. A positive critique of the welfare state would aim to restructure welfare institutions so as to bring them more into line with a detraditionalized world of manufactured uncertainty. Many interesting and important issues are raised by such a reorientation, although this is not the place for such a discussion.

Positive welfare means the active mobilization of life decisions rather than the passive calculation of risk. We should think in terms of positive welfare, not only when considering the position of the welfare state within the developed societies, but also when approaching the seemingly intractable problem of the divergence between the rich and poor countries globally. There is a shift in political orientations going on today which corresponds in a general way to the shifting circumstances of social life discussed thus far. This is a transition from emancipatory politics to life politics. By 'emancipatory politics' I mean the pre-given political arena of Left liberal political theory and practice. Emancipatory politics is concerned with securing freedom from oppression, with social justice and with the diminishing of socio-economic inequalities. It has also been the defining parameter for Conservatism; Conservatism arose as a reaction precisely to the Left liberal values held first of all in the American and French Revolutions.

Emancipatory politics is a politics of life chances. The relevance of emancipatory political problems does not diminish with the advent of life politics; instead, life political issues come to form a new set of contexts of political decision-making. Rather than a politics of life chances, life politics is a politics of life decisions. It comes to the fore in the degree to which the end of tradition combines with the end of nature. In many areas of social life thus detraditionalized, new decisions have to be taken; these decisions are almost always politicized, involving as they do an ethical or value dimension. Crucially, however, issues of life politics cannot be settled by emancipatory political criteria.

The debate surrounding abortion is one example of a life-political issue. Where abortion becomes both easy to obtain and non-dangerous, a whole series of novel questions are posed. The issues involved in the abortion controversy, however, do not conform simply and directly to questions of emancipatory politics. The women's movement raised the right to easily available abortion as an emancipatory issue, but the problems posed by abortion cannot be resolved by such means alone: they concern questions such as 'at what point is the foetus a human being?'.

A second illustration of the emerging agenda of life politics is the controversy over the family. In most countries the family has suddenly become politicized, and the discussion of 'family values' intensified. Why should this be? The answer lies in the detraditionalizing of family life, something happening not only in Western countries. The discussion going on about the family certainly continues to raise issues of emancipation, but is by no means limited to them. Many issues are raised which are connected instead with the ethics of life decisions. The family is no longer equivalent to a state of nature, but rather is being reconstructed afresh.

The more that life-political questions move to the centre of the political agenda, the more it makes sense to think of the emergence of a 'post-scarcity society', particularly within the industrialized countries but to some extent across the world as a whole. The idea of a post-scarcity society has a lengthy history, and it is important to distinguish my usage of the term here from others that have been adopted. One sense of the term 'post-scarcity' surfaced in early socialism and also found expression in Marx's youthful writings. In this sense, 'post-scarcity' meant the universalizing of abundance. Marx at least hinted at the possibility that industrial society could create so much wealth that everyone might have enough to fulfil all possible needs. Scarcity would more or less disappear. This is not what I mean by the notion. Some goods, including especially positional goods, will always be in short supply;

and the world being as it is, there seems no chance of the creation of a social order of super-abundance.

In more recent years the idea of post-scarcity has quite often been linked to the so-called 'Inglehart thesis'. On the basis of survey evidence, Ronald Inglehart (1977) has proposed that a current of 'post-materialism' is moving through the industrialized countries. People are turning away from the overriding goal of economic growth and orienting their lives towards different values. In so far as it is valid, the Inglehart thesis is certainly relevant to the notion of a post-scarcity society as I use it, but does not offer an exhaustive characterization of the term.

By a post-scarcity society, I mean not a distinctive form of social order, but a series of emergent trends. These trends are the following:

- The increasing involvement of political debate with questions of life politics.
- The diffusion of circumstances of manufactured risk from which no one can be completely free. Some, but not all, ecological risks are of this type, although ecological hazards are only one form of generalized risk.
- A decline in 'productivism', where this term is taken to refer to a pre-eminent commitment to economic growth. Productivism sees paid work as the core defining feature of social life. It is this aspect of a post-scarcity society which most closely overlaps with Inglehart's formulations.
- The growing recognition that the problems of modernity cannot necessarily be resolved through more modernity. This refers in effect to a broad consciousness of the importance of manufactured uncertainty. Many examples can be found in the area of technology and technological innovation. The impact and value of technological innovation cannot be decided solely in technological terms. For instance, no amount of technical information will show conclusively whether or not a nuclear power plant should be built; such a decision involves an irreducible political element.

In so far as tendencies towards the formation of a post-scarcity society do in fact develop, they are likely to alter the conditions of socio-economic and political bargaining, both within and across societies. There are some positive implications here for issues of poverty and inequality. Grasping these means indicating the relevance of certain kinds of life-political questions for more well-established issues of political emancipation.

Existing prescriptions to do with alleviating inequality tend to be based upon possibilities of the direct transfer of wealth or income from more affluent to poorer groups. I do not suggest that attempts to provide such direct transfers should be abandoned. They have distinct limitations, however, especially in so far as they are bound up with difficulties of the welfare state noted previously. There are some interesting similarities between the critiques of the welfare state which have come from the political Right and critiques of welfare aid programmes internationally, most of which have come from the political Left. In the context of the welfare state, Rightist authors have argued that, for example, the building of large housing estates creates more problems that it resolves. Such estates destroy pre-existing modes of communal life and foster welfare dependency. Those on the Left tend to resist such analysis when applied to welfare institutions, but present a quite similar argument when discussing the drawbacks of global aid programmes. Where such aid is used, for instance, to build a large dam, critics argue the result is often the displacement of local forms of interdependence and the creation of new forms of dependency upon the bureaucratic provision of resources.

Thinking laterally about alleviating inequality makes it possible, at least in principle, to escape from such dilemmas. Instead of thinking primarily in terms of direct wealth or income transfers, we should consider the possibilities implied in what might be termed 'life-style bargaining'. Life-style bargaining involves the establishing of trade-offs of resources, based upon life-political coalitions between different groups. Four main types of life-style bargaining may be distinguished. Each can, in some circumstances, be redistributive downwards, although I would stress that in each of these contexts opposite possibilities also exist.

The first form of life-style bargaining depends upon active risk management. There are many situations, both within and outside the developed countries, in which the active management of manufactured risk can generate a positive redistribution of resources. An illustration can be taken from the area of health care. There is normally a quite direct correlation between poverty, both relative and absolute, and the risks of contracting various kinds of illnesses. It is not always the actual condition of poverty itself which produces this connection; rather, the connection comes from certain life-style practices which those in poorer groups tend to follow. Programmes of health education, diet and physical self-care can quite readily be redistributive downwards. Those who benefit most from such programmes tend to be people in poorer groups, who ordinarily do not have the same access to relevant information and strategies as do more affluent individuals.

A second type of life-style bargaining is economic life-style bargaining. In this case there are direct economic trade-offs between groups. A major area of economic life-style bargaining concerns the distribution and nature of paid work. There are powerful trends tending to accentuate inequalities in the domain of work. Some have argued, for example, that a generalized lowering of wages of workers in less skilled jobs is occurring, because of the impact of global competition — firms have an interest in reducing the costs of labour wherever possible. Moreover, it may be that new technology will eliminate jobs without the creation of new demand which would generate jobs to replace them.

Yet not all changes affecting paid work have such negative implications for equality, and it is readily possible to point to trends and active policies which could move in an opposite direction. In a world where the amount of available work may shrink substantially over the coming twenty years, the distribution of work holds the key to overall social integration. I list here only an example of a situation in which life-style bargaining over work can be redistributive downwards. There is a tendency for people (particularly men) in well-paid jobs to retire much earlier than they used to. Some such early retirement, of course, is involuntary, and the jobs which individuals lose in that case are not necessarily replaced — at least by work of a comparable level. The larger proportion of such early retirement, however, is deliberately chosen. These are people who become 'time pioneers', people who regard the flexible control of their careers as more important than a strict work orientation. In leaving jobs which they could have held on to, they release them for those of a younger generation — with a 'chain of opportunity' effect down the line. The work thereby redistributed may 'filter down' in a patterned way, not altering the distribution of income and wealth very much. Yet if a single job is thus created for a young person, even if that job is relatively poorly-paid, the result is likely to be a downwards redistribution of resources, since younger people are disproportionately represented among the 'new poor' and among the unemployed.

A third type of life-style bargaining is ecological. As with the other categories, we know that ecological objectives often clash with attempts to produce a downward distribution of resources. Ecologically-sensitive policies are sometimes expensive, and may go against the economic interests of power groups. For instance, regulations aimed at limiting industrial pollution can run counter to maintaining forms of industrial production which generate employment for poorer people. The ecological news, however, is not by any means all bad — there are many circumstances in which ecological life-style bargaining can be redistributive downwards. This applies both within the developed societies and in more global contexts. The reason is that poorer people, by the very nature of their circumstances, are often forced to adopt life-style practices which are ecologically damaging. Such is the case in instances ranging from fuel pollution in the developed countries to the cutting back of rain forests in impoverished Third World areas.

As in the other areas of life-style bargaining, there is a diversity of contexts in which more affluent groups share an interest in reducing such ecologically harmful practices. As a minor example, take the policy which has recently been introduced by some European governments of paying a sum of money to the owners of vehicles which are particularly polluting if they trade in those vehicles for newer, less environmentally harmful ones. Since poorer people tend to be the owners of older vehicles, which emit more damaging emissions, this type of policy tends to be redistributive downwards.

The fourth form of life-style bargaining might on the face of things seem much less important than the others, in so far as material inequalities are concerned. This is what I shall describe as emotional life-style bargaining. Far from being the least important type, however, it is in some ways the key to all the others. It refers to negotiation about the emotional conditions of our lives, and these conditions have changed as massively as any of the more formal contexts of social activity in response to the wide social transformations described earlier in this paper. Particularly important here are the changing relations between the sexes, a phenomenon of worldwide importance and certainly not limited to the economically advanced societies.

Women across the world now stake a claim to forms of autonomy previously denied or unavailable to them. Such a claim plainly has a strong emancipatory element, in so far as a struggle is involved to achieve equal economic and political rights with men. At the same time, however, that claim to autonomy intrudes deeply into the domain of life politics, for it raises issues to do with the very definition of what it is to be a woman, and therefore a man, in detraditionalizing societies and cultures. Few things can be more significant worldwide than the possibility of a new social contract between women and men, since sexual divisions affect so many other forms of stratification in societies of all types.

To the extent that it could be achieved, a new social contract between the sexes would certainly be redistributive downwards. For women are everywhere on average less privileged than men, and again make up a disproportionate part of the 'new poor'. Redefinitions of gender and sexuality rebound directly, not only upon the sphere of the family, but upon that of work. Most innovations or changes which improve the working conditions of women reflect back on other inequalities — and the reverse is also true. And what of men? Suppose it became increasingly common for men to redefine the emotional and communicative balance of their lives, moving away from the primacy of paid work and other activities in the public domain. Many consequences tending towards greater economic equality would stem from such circumstances, ramifying through most contexts of social life.

This discussion of positive life-style bargaining may sound utopian, given the strength of the influences tending to produce large-scale inequality, social division and even social fragmentation. It should be stressed that I am not suggesting that there is any inevitability about the downward redistributive effects of life-style bargaining. Yet whether we like it or not, in conditions of manufactured uncertainty and detraditionalization such bargaining is likely to become a central feature of formal and less formal political manoeuvring. Within the developed societies, a variety of new pacts, some of which will figure directly in electoral politics, are likely to emerge in the future. One such pact, for instance, might be between older people and the young, for both figure among the more deprived groups in the contemporary world. As always, the currents affecting social life do not have an inexorable character. We always have possibilities of individual and collective choice — this is the very core of life politics in any case. We can try to use whatever choices we have in a fruitful way. Life political mechanisms offer us the possibility of defending some of the emancipatory values which otherwise, paradoxically, are likely to lose their purchase.

REFERENCES

Beck, Ulrich, Anthony Giddens and Scott Lash (1994) *Reflexive Modernisation*. Cambridge: Polity Press.
Inglehart, Ronald (1977) *The Silent Revolution. Changing Values and Political Styles among Western Publics*. Princeton, NJ: Princeton University Press.
Leach, Edmund (1968) *A Runaway World?* London: BBC Publications.

On the Social Costs of Modernization: Social Disintegration, Atomie/Anomie and Social Development

Johan Galtung

1. THREE THESES ON SOCIAL DISINTEGRATION

The first of my three theses reads as follows: many human societies (perhaps most) are in a state of advanced social disintegration at the close of the twentieth century — at the threshold of the third millennium AD. This does not mean the situation is irreparable. But it does mean that remedies have to be found and enacted quickly, partly to halt disintegration (negative social development) and partly to build more solid societies, not only integrated but less susceptible to social disintegration (positive social development). Such societies should also be capable of providing 'human security', here interpreted as satisfying basic human needs (positive human development), or at least of reversing processes of human needs degradation (negative human development). In the same vein, they should be capable of enhancing the ecosystem,[1] building diversity and symbiosis (positive nature development), or at least of halting processes of ecosystem degradation (negative

---

1. As used here, the 'ecosystem', or nature, includes the homosphere (humans and the 'man-made environment'). The word 'environment' is avoided because it draws a misleading line between the human and the non-human.

nature development). To this should be added a world dimension: if the world is a society of societies, that society should also be integrated (positive world development), or processes toward disintegration (negative world development) should be reversed.

We thus have four spaces of development (Nature, Human, Society, World) and for each one a more modest negative task and a very ambitious positive task: a tall bill indeed! In addition, these lofty goals may not even be compatible: a disintegrating society may be more flexible, capable of meeting new challenges; and an integrated society may be too rigid to take on new tasks creatively. That, however, remains to be explored.

Dramatic, somewhat apocalyptic statements like the thesis above are frequently heard nowadays. They can be brushed aside as more cases of 'drama supply' to meet a perennial 'drama demand'. Another, less reassuring, interpretation would be that there might be much truth to them. One point should be made here: a thesis about social disintegration is not, in and by itself, a statement about eco-crisis (depletion, pollution, over-population or any combination of the three), about misery, unemployment, low or negative economic growth, or violence and war. The statement is about society as something *sui generis*, of its own kind, as sociologists have always insisted.[2] 'Social disintegration' is an additional problem, closely related to and perhaps even more significant in its consequences than all the other global problems included under the headings of nature, human and world development. Being different, the problem will hardly yield to remedies designed for the old problems: new approaches are called for.

---

2. Thus, sociology is not aggregate psychology, nor aggregate social psychology. The conceptual building bloc of sociology, the atom so to speak, would be interaction, the interact, between at least two actors. The sum total of zillions of patterned interacts constitutes a *structure*, a molecule of interacts often of high complexity (like a protein molecule in chemistry, one reason why organic chemistry may be a useful metaphor for sociology; see Johan Galtung 'Structural Analysis and Chemical Models', Ch. 6 in Galtung, 1978: 160–89). These interacts have to be filled with concrete human beings (at least in a human society), and with human beings come their personalities with their various layers shaping their concrete life with themselves (psychology) and with others (social psychology). Society may generate such structures according to some common mould (like pyramid structures, wheel structures), referred to as 'deep structures'. Human beings are steered, to a large extent, by the structures in which they are embedded. But they are also steered by their *culture*, the symbolic, meaning-giving aspect of the human condition, particularly by the normative, valuative part. Values may be conscious or subconscious, and individually or collectively held. The category 'collective subconscious' may be referred to as 'deep culture', seen here as generating conscious and/or individual values — but always with significant variations. In sociology as a science, 'structure' is more emphasized than 'culture', as any cursory perusal of a journal of sociology will show. But structural analysis without culture reduces human beings to robots, programmed with no consciousness of their programming and no access to changing their programmes; and cultural analysis without structure elevates human beings to a freedom which is not ours.

Figure 1

Space	Global Problem
Nature	*ecological degradation, population*
Human	*poverty/misery, repression,* spiritual alienation
Society	*economic underdevelopment,* social disintegration
World	*massive violence, war (inter-state/intra-state)*
Time	non-sustainability
Culture	inadequacy

So let us identify social disintegration as a 'global problem',[3] among other global problems, distributed on the spaces of the human condition used above, adding the 'spaces' of time and culture. The problems italicized in the figure have already received general attention to the point of being the basic foci of the many endeavours by the United Nations under the headings 'environment' (for nature), 'human rights' and 'development' (for society) and 'peace' (for the world). A time dimension has been added recently: 'sustainability'. Although nobody is in favour of non-sustainable solutions to the problems of environment, human rights, development and peace, this is a useful reminder of the importance of solutions being reproducible, if possible even self-reproducible (as opposed to stop-gap measures or measures that consume more problem-solving resources than they produce).

The other three problems on the list above have not yet entered the general discourse. There are reasons for that. The specialists on 'spiritual alienation' would be religionists and psychologists; on 'social disintegration', social scientists in general and sociologists in particular; and on the possible 'inadequacy' of mainstream (meaning Western) culture, religionists again, cultural anthropologists, philosophers. In other words, they require new expertise — so far mainly limited to UNESCO meetings. These concerns do not carry the same weight as the natural sciences, economics and security studies, which are assumed adequate for the problems discussed.

The three additional problems are also found at the core of the dominant social formation, in and of the West. They imply questioning individual internalization, social institutionalization and culture. Lives lived without

---

3. The term 'global problem' is very frequently used, probably to call the attention (of policy-makers, of people with money) to the gravity of the problem. Three uses can be distinguished: 'global' in the sense of 'worldwide', being shared by a high number of societies; 'global' in the sense of 'world-interconnected', with causal loops spanning the whole world; 'global' in the sense of 'world-system', applying to world society as such. Social disintegration as a 'global problem' would cover all three uses.

meaning, societies disintegrating, cultures without answers are serious problems *sui generis*; not only side-effects or side-causes of the problems of eco-breakdown, misery and war. Moreover, all of these are strongly related.

For the second thesis we need a simple definition formula: social = structural + cultural. By 'structure' we simply refer to 'patterned interaction', the macro, gross, general picture of 'who relates to whom, how, when and where'. This is social traffic as seen from the top of the Empire State Building, not by watching drivers from the corners of Fifth Avenue and 42nd Street in New York City. The key word is pattern, not the individual variations. There are no individual name tags. Human beings appear as 'driver', 'cop', 'pedestrian'. The structure changes over time. The term is inseparable from the term 'process'; there may be stability, secular trends up or down, cycles (with any period, like the cycles of 24 hours and 365 days).

By 'culture' we mean the 'what' and 'why' of interaction; and the 'what not/why not' that is important in explaining missing interaction: the structure not there, the absent link of interaction. Whereas interaction is *between* actors (and patterned interaction is the mega-version of the single inter-act), culture is *within* actors. It may, however, be shared: patterned culture is the mega-version of the individual why and why not; the mutual rights and obligations of interaction, the expectations, or binding normative culture.

The second thesis can now be formulated: at the roots of social disintegration is a twin process of destructuration and deculturation, heading for structurelessness and culturelessness. Following Durkheim we shall refer to culturelessness as 'anomie',[4] and then introduce a neologism for structurelessness, 'atomie'. Of course we have not come that far. Society is not yet a heap of mutually isolated social atoms, individuals; and there is still much binding normative culture around. But we may be on the way.

On the way to where, to what? To a society of Leibniz's monads,[5] fully self-sufficient? Obviously not, for human individuals can hardly survive in

---

4. Independently of the present paper, Durkheim's concept of anomie (also used extensively by Merton) has recently been taken up by the Swiss Academy for Development as a major research focus, inspired by the World Summit for Social Development. See Swiss Academy for Development (1995).

5. Leibniz's 'monadology' constructs human beings not as individuals in interaction, but as basically self-sufficient units, in need of no help from others. The monads are without windows since there is no need to engage in what sociologists refer to as 'complementary role-expectations' (I expect you to do A if I do B; and since I want you to do A, I'll do B). The strong unit is in no need of others. This all promotes the best of all conceivable worlds because of a pre-established harmony (by God). The monads will not collide; they are steered, not by mutual rights and obligations, but from above, like driverless cars or planes on pre-programmed autopilots in need of no human intervention; very different from Leibniz's contemporary, Spinoza, to whom friendship was a basic concept, and also from Martin Luther's focus on the strong individual, *hier steh'ich, ich kann nicht anders* (Leibniz

total isolation.[6] But we can easily imagine inter-action reduced to a thin minimum, like some e-mail contact; making society a set of isolates[7] more than a structure relating positions filled with individuals. In other words, the actor would be the isolated individual as such, not the individual as, for instance, 'head' of the family, CEO (Chief Executive Officer) or SEO (State Executive Officer, the head of state/government). And the normative culture informing these individuals about what to do would be centred on that which serves the individual. No interacts, only acts.

In short: at the end of the road winding through history and into the future we see a social formation ('society' may no longer be the term) basically atomized into individuals, thinly and weakly related, each acting out of egocentric cost-benefit calculations. We are close to this state of atomie, but there is still some interaction left. We are also close to anomie, where the only binding normative culture left would be individualized cost-benefit analysis. Anarchy would be another term, *bellum omnium contra omnes, homo homini*

---

was a Catholic). I am indebted to the Norwegian philosopher Arne Næss, in a private communication, for this interpretation of Leibniz. The Scandinavian will be reminded of Strindberg's *ködets lust och själens obotliga ensamhet* (the lust of the flesh and the incurable loneliness of the spirit), although Leibniz seems mainly to pick up the latter point.

Two other metaphors may be useful as well. There is the case of the noble gases in chemistry (argon, helium, neon, krypton, xenon, radon), which are said to be 'noble' because, since they are not ionized, they do not connect with any other elements to form compounds. A set of monads, or hermits, would be like a noble gas with no compounds; they would certainly be nothing like a protein — a compound of compounds (amino acids) that might serve as an image of a society (the amino acid being, for instance, the family). Then there is the idea of the human being as a world of cells, connecting colonies of cells of the same kind. Cells relate, they communicate, e.g. through sodium channels. They even relate by committing suicide, offering themselves up to each other (apoteosis). And they die (necrosis). If they did not relate to each other, then the body would die; the whole body derives life from the change within and the exchange between cells. We use many expressions drawn from the metaphor — *le tissu social*, for example.

6. Some recent data on the size of households may be of interest in this connection. Thus, 40 per cent of the households in Sweden consist of one person. Often this one person, very likely an elderly woman, will live in an 'apart-ment', somewhat akin to a Leibnizian monad, with windows not to other people as persons, but to a 'view'. Of course, a single person in an apartment is not self-sufficient. But if the paycheck from the welfare state, the neighbourhood supermarket and the urban services (such as water, electricity and sewage) are included, then we are close to the monad. The harmony may not be divine, but established by the joint action of State and Capital, with the Media thrown in for the human mind — again with no need to meet anybody in direct human interaction.

7. I do not say 'individuals' on purpose, as that term is meaningful in a social context emphasizing individual differences; and differences can only come to the fore in a social context of interaction. Individualism, in the sense of self-differentiation from others and self-assertion, is only meaningful when there are others around to compare with or to interact with. Collectivism differs from this only at one point: the 'self' above is a collective Self, such as 'my group' or 'both of us'. Isolates neither differentiate nor assert themselves this way or that; the social context is irrelevant.

lupus. The social fabric (*le tissu, el tejido*), the social body, *lo social*, falls apart.

The third thesis might read something like this: we are at a stage in human history where the problem is not only whether interaction structures between individuals, groups and countries are right or wrong, but whether there is any structure at all; and not only whether the culture defining right or wrong is right or wrong, but whether there is any normative culture at all.

On the road we would expect a number of social phenomena. First, we would expect the focus of interaction to shift from 'mutual rights and obligations', a reciprocal mix of egoistic and altruistic orientation, to an egoistic orientation of 'what is in it for me'. For organization members the shift is from reciprocity to 'what can the organization do for me'. Like predators they descend upon macro-organizations like State and Capital, preying on them for individual benefit, then withdrawing with the booty. Meso-organizations like NGOs, including parties, trade unions and churches, are used as stepping stones. Micro-organizations, like families and friends, are not spared. Spouses will demand services like sex and security, and in addition 'freedom' (particularly husbands). The offspring see the family as a launching platform in life and offer little or nothing in return after — and even before — take-off.

Second, we would expect increasing corruption at all levels of social organization.[8] By 'corruption' we mean a way of using organizations for egoistic purposes, influencing decisions by injecting resources (money, sex) into the process; corruptor or corruptee acting out of egoistic cost-benefit analyses. Third, with social nets (organizations) decreasing in significance, and social knots (individuals) on the increase,[9] we would expect increasing mobility out of nets, relations and organizations, indicating that they have been used. After exit there may be entry into new ones, or into individual monads. People will vacate bonds between spouses, parents and children, siblings, friends, neighbours and colleagues, frequently and easily. New relations may become increasingly thin, shallow.

Fourth, we would expect increasing violence at all levels of social organization. There would be no absolute, binding norms standing in the way, no *homo res sacra hominibus*. Other human beings inside the organizations will be seen as substitutable — the relationship being so thin anyhow — and hence as expendable. Outside the organizations they will be seen as resources. The utility supposedly accruing from violent acts will be weighed against the disutility of punishment and the probability of detection/

---

8. Transparency International (TI), in Berlin, is the organization concerned with corruption, as Amnesty International is concerned with human rights in general, and political prisoners and torture in particular. For an analysis of corruption from a TI point of view see F. Galtung (1994).

9. The net-knot metaphor is taken from the writings of the Indian-Catalunyan-American philosopher Raimundo Panikkar.

punishment. As violence becomes pandemic, the latter probability will tend to zero given the asymmetry between the ease of committing a crime and the difficulty of detecting it. Fifth, we would expect increasing mental disorder, assuming that human beings are not made for high levels of atomie/anomie but for interactive human togetherness, guided by mutual rights and obligations, in thin *and* thick human relations, definitely including the latter. Types of conduct indicative of mental disorders, such as drug consumption, alcoholism, sexoholic and workaholic behaviour, perverse physical and verbal violence, are also efforts to find identity in tighter and thicker human interaction and in the deeper recesses of the Self. They are outer and inner journeys. When such efforts fail, suicide is a possible way out; not only out of despair, but also as the ultimate act of egoism.

Summarizing, this is a fairly bleak — some would say far too dark — image of human society today. But the problem right now is to understand these processes in order to arrive at some idea of where we are right now; *où en sommes-nous*. For that, we need some kind of macro-historical perspective, with all the shortcomings of abstracting and generalizing from a super-complex reality.

2. A MACRO-HISTORICAL PERSPECTIVE: STRUCTURAL TRANSFORMATIONS

Imagine that we now divide human history into four phases, calling them 'primitive', 'traditional', 'modern' and 'post-modern'.[10] In other words, 'modernity' is not seen as the end of history and certainly not as global market economy *cum* democratic polity — a social formation seen here as highly unstable. The fourth phase, the phase that comes after modernity, as the Middle Ages come between antiquity and modernity and 'metaphysics' comes above or after physics, is the post-modern phase. The term is frequently used; the following is an effort to give it a richer connotation.

'Primitive' will be identified with mobile hunter-gatherers and nomadic pastoralists; 'traditional' with sedentary, local agriculture and the emergence of classes and castes that do not have to engage in manual work for a living; 'modern' with the large-scale organizations of State, Capital and Media, building state, regional and world bureaucracies, markets and meanings; and 'post-modern' with the destructuration and deculturation alluded to above. The post-modern society is seen as essentially chaotic and anarchic for reasons to be given in more detail below. In other words, it is not seen as a global version of modernity but as its antithesis, or as one of several antitheses.

---

10. For my own first effort to do so, see Galtung (1980). In that paper, from 1967, I used the term 'neo-modern' not 'post-modern', since I think it is 'modernity' extrapolated rather than some new paradigm. However, the term 'post-modern' is now so frequently used that it is hard to avoid.

Reduced to a brutally simplistic formula, this is the story of humanity on its way from nomads to monadism. For that social story to be told we shall proceed on the two parallel tracks above; one structural and one cultural. To do this some concepts are indispensable. Some references have already been made above to *thick vs. thin* interactive relations. Let us now shift to *primary and secondary* relations, defining primary (in the Weber-Tönnies-Sorokin-Parsons tradition) as 'diffuse' ('thick') and particularistic, meaning relating to that particular Other, not to anyone of the same kind (in other words, the relation is non-substitutable). The definition of 'secondary' would be based on the opposite pair: 'specific' ('thin') and universalistic, meaning treating everybody of the same kind, who embodies the same (low) number of characteristics, the same way. The classical examples of primary relations would involve close relatives, the more remote (cousins four or more times removed, for instance) being treated the same; and friends; and enemies. But it would also include colleagues and neighbours, work places and voluntary organizations. In short, kinship and friendship, vicinity (also community) and affinity, workship (also school) and worship. High interaction frequencies will rub off; over time small-and-thin relations will be thicker and less standardized. For all six cases some collective Self is defined, offering identity and some security in return for some altruism.

Let us then introduce another variable, so often missing in social analysis: *size*, the sheer number of people involved. Let us divide organizations into 'small' and 'big', the dividing line being roughly the upper limit to the number of people a human being can identify, and relate to, positively and negatively.[11] The order of magnitude would be 10^2 to 10^3. Since primary

---

11. If we have N actors (persons in a social system, societies in a world system), then the minimum number of links needed to connect them is $N-1$. Each actor, except those at the end, is connected to two others in a hierarchy with an apex, connecting downwards to two others, or in a chain. There is no limit to size; the pyramid can have any number of layers, the chain is endless and may even become a circle. But if the rule is that each actor shall relate to everybody else, then everybody with no exception (there is no longer anybody 'at the end') has $N-1$ links to manage and the total number of links is $N(N-1)/2$, in other words N/2 times more:

N	Alpha $N-1$	Beta $N(N-1)/2$
2	1	1
3	2	3
4	3	6
5	4	10
6	5	15
7	6	21
8	7	28
9	8	36
10	9	45

With increasing size there is no load increase for each actor in Alpha and the total interaction increase is very slow. For each actor in Beta there is an increase (like relating to 9 friends in a group of 10) and the total interaction increase is much quicker. Obviously, there is an upper limit to how much interaction a human being can handle.

relations are based on identification, we arrive at the simple conclusion that big-and-thick is impossible. Secondary relations will tend to be big (and vice versa); only when small can they be primary.

Thus human interaction structures come in two basic modes: thin-and-big and thick-and-small. Let us call them 'Alpha', the pyramid, and 'Beta', the wheel (for a first publication on this, see Galtung, 1979). In modern societies Alpha is organized by the three pillars of society, State, Capital and Civil Society, in the form of huge bureaucracies (including armies and universities), corporations and people's organizations. But inside Alpha, small informal Beta structures of people with primary relations, such as colleagues who become friends, or enemies, would be nesting; growing in cafeterias, over repeated encounters in lifts, some evolving into super-Beta relations known as love. Seen from Alpha they all introduce personal and subjective elements into the impersonal, objective atmosphere of a perfectly constructed Alpha, with everybody substitutable, even if this means alienated.[12] Alpha people are right: those who spy on Alpha centres for state and corporate secrets often use Beta networks, including love relations, to get access, like the classical secretary making extra photocopies for a friend (as one example, see Haase, 1993).

Let us then introduce a third variable, vertical versus horizontal, here seen as relational, not only relative, and as exploitative — grossly asymmetric in terms of net benefits.[13] Why do people enter such vertical, exploitative relations? Because, forced by coercion or tradition, they may have no choice. The alternative to exploitation may be starvation (Marx on capitalism). The result is vast action spaces for people on top, strait-jackets at the bottom; material enrichment on top, impoverishment lower down.[14] Challenges on top, routines at the bottom. In horizontal relations this is better distributed; gross asymmetries lead to break-ups in thin relations.[15]

Alpha tends to be vertical. Layer can be added to layer, in principle covering all of humankind through processes of globalization in one big pyramid or hierarchy with a single apex. This projection of the State would be known as World Government and the corresponding projection of Capital as the World Market. The present G-7 has aspects of both. But so far

---

12. The argument here is precisely that substitutability is the structural basis for alienation, making people detachable from their work product, from others and ultimately from themselves — like anything else in throw-away society.
13. This theme has also been conspicuously absent from many paradigms in US sociology.
14. Another metaphor: a normative, iron strait-jacket at the bottom, a rubber suit to grow and expand in at the top.
15. Two implications of this are very well known. First, in order to make break-ups (such as strikes, lock-outs, firing people, making them 'redundant') less costly in human terms, relations are thin. A social distance is kept between employees and employers. Second, if relations are thick (as in a marriage), break-up becomes extremely time-consuming and costly in human terms. Thus one obvious strategy to protect oneself against such costs is to keep relations thin, as in 'affairs', a term hinting at business-type relationships.

Alpha-ization is clearly more pronounced at the regional than at the world level, the European Union — as seen in the Maastricht Treaty — being one example. (The Soviet Union was another, but State and Capital were more clearly merged into one pyramid than in the European Union or the United States.)

Beta can be both horizontal and vertical ('Gamma'). A tribe run by chiefs and shamans, villages run by Big Men and land-owning families, families run by a *pater familias*, marital relations under conditions of patriarchy (and the infrequently found matriarchies), or the small farm/firm with very tight and very authoritarian relations under the 'boss', are thick and small and also vertical. They can be horizontal, as in kinship and friendship/enmity groups, among neighbours and colleagues; with other human beings in general, in worship and workship.

Horizontal Alpha structures can also be imagined ('Delta').[16] At present electronic communication, like Internet, may serve as an example, as long as the information superhighway has a topography without centres and peripheries. Transportation superhighways tend to be rooted in big urban centres reaching into the peripheries. However, peripheries could be connected, levelling the centre-periphery gradients. In the same vein, the information superhighway will probably develop steep gradients (like toll gates) — and we are back to traditional Alpha.

As pointed out repeatedly, societies, or social formations more generally, as we know them, are mixes of Alpha and Beta. The question is how strongly

---

16. With thin-thick, big-small and vertical-horizontal, we get eight possibilities, even if for present analytical purposes we do not need to make use of all of them. For reference, here they are (two stable, two unstable and four highly unstable):

Alpha:	thin	big	vertical	bureaucracies, corporations
Beta:	thick	small	horizontal	friendship groups
Gamma:	thick	small	vertical	patriarchal families
Delta:	thin	big	horizontal	electronic/urban networks
	thin	small	vertical	unstable; will become big
	thick	big	horizontal	unstable; will become small
	thick	big	vertical	unstable; will become thin
	thin	small	horizontal	unstable; will become thick

The logic is simple. Alpha and Beta are the two prototypes for human interaction structures, the former a pyramid, the latter a wheel (with all points connected to each other). But Beta can also be vertical (Gamma), although the hypothesis will be that members will tend to escape, children by leaving, spouses by splitting up, or even through homicide and suicide. Alpha can become horizontal (Delta), as in vast networks, although the hypothesis will be that gradients will tend to build up.

The remaining four cases have not even been given names, as they are seen as highly unstable, tending toward the two prototypes. With two out of three characteristics shared with the prototypes, the hypothesis is simply that the third will have to yield, through processes like adding layers; shedding some members; making interaction more specific, less diffuse. Examples of such processes abound.

Figure 2. Human social (trans)formations: Structural macro-history

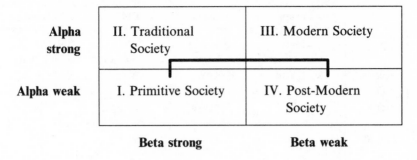

either one is articulated. Using Fig. 2, let us answer that question in terms of 'strong' and 'weak' for both, giving four combinations. (Alpha/Beta is not a dichotomy.)

Figure 2 is to be read by following the thick line which starts in the bottom left box: a humanity divided into small mobile groups, clans, lineages, small enough to be 'in-groups', with primary relations dominating, essentially kinship.[17] A tight net of mutual rights and obligations is spun inside the group, possibly with negative or no relations at all toward the out-groups they would encounter on their wanderings. They, precisely *they*, would probably be conceived of as categories of people, not even with the *differentiae specificae* given to them by Alpha logic in terms of their social positions and their qualifications for being allocated to such positions. The in-group would be too small to develop layers of verticality beyond gender/ generation and for that reason it would be well integrated socially and humanly. The weak point would be not only the thin or empty relation to other groups, but also that integration may be too tight, 'suffocating'.

With sedentary ways of producing for a livelihood and a higher level of agricultural productivity — one family working on the land producing enough surplus for 1.1, even 1.25, families — the material basis was laid for the classical caste systems (see Fig. 3). The history of traditional society becomes to a large extent the history of the relative power of the upper layers in what has to be an Alpha structure, unless the unit (e.g., the village) is small. One possibility is the ranking order indicated in Fig. 3; with the European and Indian systems being quite similar, and the Chinese and Japanese also quite similar (thus formulas like 'Indo-European' and 'Sino-Japanese' apply not only to languages). Another possibility, as pointed out by Sarkar, is a circulation of castes, in the order Kshatriya-Brahmin-Vaishya-Shudra (the Kshatriya enter to create order after the people have

---

17. Robert Redfield's 1953 classic, *The Primitive World and Its Transformations* (particularly chapter 1), and the books by V. Gordon Childe (e.g. Childe, 1946) provide fascinating images.

Figure 3. Castes: Four systems

	Europe	India	China	Japan
First	clergy	Brahmin clergy	Shi'h bureaucrats, intellectuals	Shi (samurai)
Second	aristocracy	Kshatriya warriors	Nung farmers	No
Third	merchants	Vaishya merchants	Kung artisans	Ko
Fourth	workers	Shudra workers	Shang merchants	Sho
Fifth	outcasts women children	outcasts women children	outcasts women children	outcasts women children

had their say, but they are culturally so primitive that the Brahmins enter to restore culture, but they are economically so amateurish that the Vaishya have to put the economy in order, but they are so exploitative that the Shudra people make revolts and so on . . .).

At this point solid vertical distinctions between people and elites have emerged. Alpha structures, mainly local, are being articulated. Modernity brings that process further in Alpha *strictu sensu*: country-wide, hierarchical, with a well-defined specificity in social relations stipulated in written contracts and a universalism opening the positions in the structure for citizens satisfying well-defined, explicit qualifiers. Diffuse, particularistic relations have to be weeded out from the Alpha garden, ultimately to look like the orderly French gardens that emerged at about the same time. For Beta relations, please use time after working hours and weekends.

As Alpha becomes more dominant, Beta not only becomes recessive, but starts disintegrating. One reason is simple: individual time budgets. Alpha requires full attention, because the jobs provided by Alpha are full time jobs and because the occupants of Alpha positions are not supposed to think Beta thoughts. Some Beta structures have to go, starting with such old work structures as extended families and traditional villages.

Cities are to Alpha what villages are to Beta: liberating people from the stranglehold of very tight human relations in a village, then suspending them in the thin air of urban anonymity. Cities provide more space for Beta structures than villages do for modern Alpha structures. However, these Beta structures are decreasingly related to work and increasingly to leisure, leading to the well-known pattern in many modernized countries today: villages gradually being converted from sites of agricultural production to sites of weekend leisure, and to some primary and tertiary production, plantation and tourism, for far-away buyers.

We now have to introduce a thesis, or rather a hypothesis, which is important for what follows: a Beta structure is natural to the point of being indispensable for human beings. Only Beta-type relations cater to the whole person and give the person a sense of belonging. This should not be confused with identity or sense of meaning of life; that can be enjoyed also in an Alpha structure, even in a non-structure (formation IV of Fig. 2). To belong is to have a home, somebody to relate to, somebody who knows more of the story than any bureaucracy can. The argument is not in favour of joint or nuclear families, different sex or same sex unions, those with or without children. The argument favours some Beta unit, thick-and-small, with more total relations.

One objection may be: if Beta is the natural structural environment, how is it possible for Alpha to expand at the expense of Beta? The answer is that Alpha has much to offer in the short run. For those on top, Alpha offers the material fruits of verticality — power, challenges. For those lower down, the gains may turn into losses, but the costs of being marginalized may be still higher. The Alpha lure is there — you are in it! — even as a peon in the post office in a village in East Bihar or as a second-speed EU member. For Alpha holds out a reward for good behaviour unknown to Beta: upward mobility, if not for you, maybe for your offspring. In Beta there is always room to improve the relation, to become a better friend, a better neighbour; but if an attraction of Beta is precisely its horizontality, then there is no way up. Nor is there any way down. There is a way out: if you do not behave. The problems, and the attractions, in Alpha are vertical. In Beta, they are horizontal: belonging versus loneliness.

One formula often used for modern society is Alpha for production, Beta for reproduction. From Alpha the work output may be considerable. In Beta human beings are repaired, maintained, sustained. Returning to Fig. 2, formation I would show high levels of stability, keeping humans intact, leaving few traces on nature as the work output is negligible and the consumption of natural resources likewise. Formation II leaves more traces. There will be monuments to the glory of the upper castes: temples (mosques, churches) for the clergy; forts for the warriors; market-places, banks and so forth for the merchants; poverty for the people; all wrapped together in cities. But even if human beings are exploited and repressed, they still belong somewhere — sustained and repaired; reproduced.

In formation III, however, production starts outstripping reproduction. The output is phenomenal. Alphas of all kinds get deeper roots and expand geographically and socially, covering ever larger territories, not only countries governed by states, but empires governed by mega-states. The production of goods/bads and (dis)services outstrips what anyone might have imagined. The Betas are disintegrating, and not only the extended family and the traditional villages. The nuclear family splits not only between husband and wife, but also between parents and children, and among siblings. Neighbourhoods break down when people move geographically too frequently to sustain relations based on vicinity. Invariably the same will apply to friendship and to affinity: neither can survive the high levels of social mobility — sideward, upward, downward — of modern society. Worship under the same God may still remain. (About God, however, see next section.)

The transition from primitive to traditional was made possible by the agricultural revolution, growing plants and breeding cattle in a relatively sedentary, basically Beta way. The transition from traditional to modern was made possible by the industrial revolution providing the goods, the scientific revolution providing the knowledge and the transportation-communication revolution extending Alpha reach.

But how about the transition from modern to post-modern? As we are talking about destructuration, anything removing human beings from direct interaction would count. A key word is 'tele'. Direct interaction is multi-sensorial; no telecommunication so far goes beyond the auditive and visual. Interaction is still there, but it is trimmed down, stripped, more naked. As anyone talking over the telephone without watching the facial expression and the body language knows, information gets lost in the process. And as anyone comparing telefax to telephone knows, the tone of voice may say more than the words. So the term 'information revolution' will not be used, not for the obvious reason that what is conveyed is often disinformation, but because of the high level of 'de-information' when so much quality is lost. Information retrieved from an encyclopedia or CD-ROM is not the same as information conveyed by a loving parent or concerned teacher (although the two obviously do not exclude each other).

Symbolic interaction via words or other symbols, whether arriving on ordinary or information highways, substitutes for direct human inter-action.[18] The term is 'symbolic revolution', from proclamation of edicts via modern media to automation-robotization. Alpha is there — but human relations are not.[19] If we take, as an image, Los Angeles, 1992: certain parts of the once magnificent city are wastelands. There are streets and buildings, even shops. But waste is piling up all over, the buildings are derelict and the

---

18. Thus there is a human content in a garage owner's shouting his orders at mechanics, sometimes from another nation; but not in a big bureaucracy where problems arrive in In-trays and are transformed in loneliness to solutions in Out-trays. To the objection that a really big bureaucracy cannot function like a small garage, the answer would take the form of two questions: 'Are you prepared to pay the costs in terms of atomie?' and 'Are you sure it has to be that big?'.

19. If we now define human existence not physiologically, in terms of a body with vital signs, nor spiritually in terms of a soul, but in terms of social networks, of the quantity and quality of human interaction, then the net conclusion of all of this is that formation IV consists of dying and dead human beings. Social death = physical death according to that formula; it is not only a forerunner of physiological death, as in the Western construction of life-cycles (the Childhood-Education-Work-Retirement [CEWR] syndrome). The only comfort is spiritual survival for those who believe in that.

shops are barricaded. More importantly, they are all disconnected from each other, there is not even a concept of neighbourhood. Nobody knows who is next door, nor do they care. People come, goods and services are peddled, they disappear. At night everything is locked up, dark, desolate. That is when the marauding gangs take over. They are the new nomads; the city-scape is their resource. Unable to survive in nature, they know how to survive as hunter-gatherers in the urban wastelands; hunting cars, gathering their contents. They are the products of formation IV, crystallized as a new formation I, preying on the wasteland, fighting rival tribes, including a police tribe engaged in the task of hunting and gathering gangs, the LAPD.[20] Strong Beta structures are re-emerging — ready for a second cycle?

There is a logic to this. Alpha has not disappeared, but has become very lean and mean, devoid of human content (thus, in Fig. 2, we are talking about 'weak' not 'zero' Alpha). There is work output, although some quality may get lost in this dehumanization process.[21] Much more disturbing is the question often raised by the ultimate stage of dehumanization: not only is the interaction symbolic rather than direct, but the receiver, and sometimes also the sender, is even a non-human, a robot. Robots do not crave Beta groups, they are custom-tailored for a high Alpha life expectancy. So the disturbing question is obvious: if robots do so much better, for what purpose do we have human beings at all?

The first answer is obvious: even if robots are better at production, humans are better at consumption; in fact, the whole purpose of the exercise is to liberate human beings from dirty and dangerous, humiliating and boring work, leaving all of that to robots so that human beings can concentrate on creative and non-programmable tasks and enjoy the fruits, as consumers. The second answer would be more reflective, taking into account that robots also have to be reproduced, sustained with energy and spare parts inputs, and perhaps also with reprogramming. The total cost-benefits, even done in the most naked economistic way, may turn out to be less obvious with the destructuration bill added in.

The third answer may point out that not so much is lost anyhow. With the symbolic revolution, not only can production be carried out in loneliness; the same applies to consumption. There is a neat isomorphism between assembly line production (in series) and bureaucratic production (in parallel), on the one hand, and a magazine circulating in an office (in series) and a family consuming TV programmes next to each other (in parallel) on the other. All

---

20. Los Angeles Police Department, of Gates/Rodney King/O. J. Simpson fame.
21. Thus the Japanese seem to divide their production (say, for example, of cars) into three phases: an artisanal production of parts in very small, Beta-type family firms; then assembly in giant, Alpha-type factories, to a large extent done by robots; and then a testing through dis-assembly and re-assembly by hand in small groups of highly experienced workers. Not only are human relations kept, but they are even to a large extent Beta, with a dehumanized, robotized segment in-between.

Figure 4. Structural dynamics of formations: Some basic factors

	Primitive	Traditional	Modern	Post-Modern
Alpha	weak	strong	strong	weak
Beta	strong	strong	weak	weak
Growth	low	high	high	low
Exploitation	low	high	high	low
Alienation	low	low	high	high

four cases are based on action (like turning nuts in assembly lines or zapping TV at home), not on interaction.

The sum total is not only Alpha but perverted Alpha. If the thesis of a human need for Beta as something natural is correct, we would now expect Beta to be sprouting. But what kind of Beta? Alpha supplies all goods and services, leaving few opportunities for green production on the side. If Alpha is dehumanized anyhow, then why not treat it as such? To whom do you feel more attachment, to your fellow corruptor/corruptee, perpetrator/victim, or to an abstract, symbolic structure?

Figure 4 summarizes some of the points made so far. But why do human beings engage in such exercises? Because the grass is greener on the other side. We seem to be fascinated with what is missing and to take what we have for granted, assuming it will remain there forever and not be eroded by the relentless search for the new. Until we end with a very bad deal indeed.

Of course, Primitive Man becomes fascinated with the growth and with the glory produced by traditional society. So, as Ibn Khaldun points out, the desert tribes knock down the gates and storm the city, sharing in the power and the glory, ultimately running it down for lack of *asabiyah*, solidarity (a premonition of the theory underlying the present paper). In the same vein Traditional Man becomes fascinated with the tremendous growth and power — with the national, regional and global reach — achieved by modern society. He no longer knocks down any gates, but he joins as a humble immigrant, at the margin of the host country Alpha structures, contributing to destructuration in both places. He came from reproduction without production and enters production without reproduction. He participates in building *The Wealth of Nations* at the expense of *The Moral Sentiments* — the point-counterpoint in Adam Smith's brilliant reflections.

3. A MACRO-HISTORICAL PERSPECTIVE: CULTURAL TRANSFORMATIONS

Let us now try the same story from a cultural point of view, focusing on binding normative culture, and particularly on the source of normative

Figure 5. Human social (trans)formations: Cultural macro-history

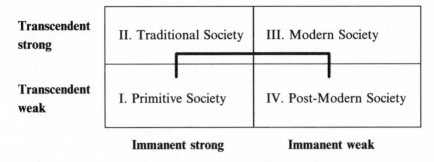

Transcendent strong	II. Traditional Society	III. Modern Society
Transcendent weak	I. Primitive Society	IV. Post-Modern Society
	Immanent strong	**Immanent weak**

culture, religion and such secular successors as national-ism, state-ism, capital-ism, science-ism. Religion contrasts the sacred and the secular: the awe-inspiring, that which cannot be touched, and the ordinary, the profane. In many religions there is also a third category: the evil, to be feared, to be avoided and, if possible, destroyed. Obviously, people are not born with, but into, a religion. There may be a basis for religious belief in the physiology of the brain (and elsewhere), but details are learned.

What, then, would correspond to Alpha and Beta? There is the theological distinction between the sacred as 'immanent', inside human beings and nature, and as 'transcendent', in a God residing outside the planet, above. That God may be a Mother God (as in Japan) or a Father God; but in the Occident (as defined by the abrahamitic religions, Judaism-Christianity-Islam), this takes the form of Father-Sky, the Father in the Sky. The opposite would be Mother-Earth: the Earth that gives birth to our livelihood, the Earth that nourishes us and ultimately receives us upon death.

Immanent religion is more horizontal, transcendent religion more vertical. Rather than dividing religions into immanent and transcendent, however, it might be more fruitful to talk about immanent and transcendent aspects of religions. In the three occidental religions the transcendent aspect is dominant; in addition there is Evil, presided over by Satan. Prayer and submission to God are the adequate approaches. In immanent religions meditation in Self and compassion with Other may play similar roles.

However, immanent religion has a dark side, tending to be particularist rather than universalist. The sacred nature of Other may apply to the in-group only, not to the out-group. The message of transcendent religions like Christianity and Islam (but not Judaism and Shinto) would be that we are all in it, all protected from above. The condition is that we submit and pray.

Using Fig. 5, then, the story would run approximately as follows: primitive society would be protected by strong in-group norms, being tight and co-operative. Out-groups may prove friendly but also may not, so any notion of the sacred would not *a priori* extend to Other. They would have to prove themselves, not by submitting to the same Father Sky, but by relating co-operatively. They become human by being accepted parts of the social network, not by any abstract human-ness (that is probably Occidental).

Traditional society might also need some transcendent deities, particularly protective of the upper layers of society and more accessible to them than to common people. Religious relations have to mirror social relations. But the social unit is still small: transcendence and immanence can be combined. Modern society is almost inconceivable without transcendent religion, sacred or secular; a *deus* in the *rex gratia dei*. There has to be an authority beyond the apex of the Alpha pyramid as there is so much power to legitimize. Father Sky supplies the authority, not Mother Earth — she is too close to everybody. And just as imperialism established the first global super-Alphas — imperial rule and trade companies — missionarism established the homologous supremacy of universal, transcendent religion. This holds for Islamic as well as for Christian imperialism.

Immanent religion was considered pagan and particularistic, standing in the way of a universal God in need of (more than willing) missionaries and colonizers to bring the message. Imperialism and transcendent religion came hand-in-hand, one as the condition for the other. Indigenous Beta and immanent religion could then be eliminated together, as pagan, archaic.[22]

A likely objection to this would be: how about the Enlightenment and secularism in general; does this picture not paint the Occident as too religious? The answer would be that Islamic colonialism/missionarism started right after the inception of Islam ($+622$) and had the foundation of the Sultanate of Delhi in $+1192$ as one crowning achievement. From there it went eastward, stopping at the southern end of the Philippines. Christian imperialism (if we disregard the Roman Empire which was not Christian in its expansionist period) started for real in the 1490s, expanding westward (Columbus) and eastward (Vasco da Gama). The pattern was set under religious auspices. Enlightenment came to Christianity much later; to Islam (perhaps) not yet. What was needed was a universal, overarching God/Allah whose commands would be binding on all believing imperial subjects.

Enlightenment and secularization (in the West) set in somewhere on the transition from formation II to formation III. The functions of universalist/ singularist religions with Chosen Peoples still had to be fulfilled; and universally valid science claiming to represent the only possible truth, with scientists of various kinds as the Chosen People, fitted the bill. Alpha construction could now be made in the name of the three modernizations (state logic, capital logic and scientific logic) rather than religion, with Ratio — rationality — as the overriding theme. The project is still on now, under the heading of 'development assistance'.

But what happened to the Church as the Alpha prototype? The role as representative on earth of the omni-present, omniscient and omnipotent

---

22.   That extermination project is still on, for instance in the Amazon, Chiapas, Guatemala.

causa sua God went to the three pillars of power in modern society: State, Capital and the Media — the carriers of state logic, capital logic and reality representation of modern society. Underlying that, a new ethos took shape: nationalism, providing large parts of the world with national statism, national capitalism and national media, along with the disequilibria to which this leads when the territories covered by state jurisdiction, capital penetration and national settlements do not coincide. Of course, to some extent posited against State and Capital is Civil Society, with a contract (rule of law/democracy/human rights) with the State and no contract, only a market-place, with Capital and Media. New priesthoods emerged as carriers of the new faiths: jurists for the State; economists for Capital; journalists for the Media; political ideologists for Civil Society and nationalists for the Nations.

In short, the structure of the transcendent God, chosen by people as object of worship, remained intact. The places of worship were different, the content of the prayers varied, but the submissiveness remained. For top positions in Alpha new faiths were needed, such as allegiance to the new priesthoods, meaning concretely faith in the human Ratio and such products as jurisprudence and mainstream economics.[23] In addition comes faith in the (virtual) reality images produced by the Media, and in nationalism. Modern society has been labouring under such formulas for some centuries now.

Thus, human beings were almost deprived of immanent religion through the missionary activities of the religions of the imperial powers. But with that project still on, the second project of the West, secularism, started undermining transcendent religion, leaving human beings deprived of Father Sky, with no Mother Earth as alternative, and only small groups (Quakers, Buddhists) still insisting on the sacred nature of life, particularly human life. This is exactly formation IV; for secularism, in the shape of humanist ethics, has not been capable of producing binding norms for human behaviour. Why shall you not commit adultery, kill, steal and lie when other humans are mere objects and there is no accountability to higher forces as there is no transcendent God anyhow?

The final result is the total anomie of formation IV, with human beings left with the only normative guidance that always survives: egocentric cost-benefit analysis. The point is not normlessness, the point is that norms are not binding. That is the meaning of culturelessness. The process has gone quite far.

---

23. Both of these are, like theology, deductively constructed — based on a number of hypotheses or rather axioms (like the Enlightenment faith in human 'rationality') that gives them an apodictic character. But since they are secular substitutes for religion, their dogmatism (even to the point of *credo quia absurdum*) is not criticized. Their basic credos (rational human beings with knowledge of the law will be law-abiding; rational human beings act so as to maximize their utilities) are unfalsifiable, meaning that no empirical evidence can be used to unseat the new high priests.

4. ATOMIE AND ANOMIE IN DOMESTIC SOCIETY:
SOME IMPLICATIONS

In Figs. 2 and 5, two processes have been indicated through four social formations. How far concrete societies, groups or individuals have come along these trajectories can only be decided through empirical studies, but one interesting point emerges: the more modernized society becomes, the further advanced it will be along this trajectory, since by 'modernization' we mean precisely the triumph of Alpha over Beta as dominant social formation, and the triumph of Ratio over the Sacred as dominant moral guidance. What was not taken into consideration was that human beings may need *both* Beta (for their personal sustenance) and the sacred (for life to have a meaning and their action to be guided). Alpha alone, and Ratio alone, have provided us with material abundance and impressive control and co-ordination structures (in need of the counterforces generated by Civil Society, though, and with the Media oscillating in their loyalties to State, Capital and Civil Society). But deep sustenance and guidance they cannot offer.

Then two things went wrong, both basically unintended. Together they catapult us into formation IV, atomie cum anomie. First, Alpha became more and more naked, stripped of human content, as Ratio provided Alpha with its many gadgets. Take only one example: automated telephone systems, not only bypassing the switchboard lady through direct dialling, but then landing the caller with 'if you want . . . , push 1', some canned music and finally a recorded response. Whether done to save labour expenses (and time), to standardize responses, or to save the recipient from any further argument, the net result is destructuration, as there is no (or very little) direct human interaction involved. Second, the hope must have been that Ratio, seen by the great Western philosophers as essentially universal, would provide a basis for a binding ethic. The problem is not that Ratio may be less universal and more a product of the general code of the many human cultures, but that Ratio does not generate sufficient ethical commitment.

At this point the synergy between the two trajectories heading for atomie and for anomie is set. Alpha, in the shape of a modern educational system, is very good at schooling people in the products of Ratio, at the level of primary, secondary and tertiary education. The two not only fit each other by being standardized up to the country and regional levels, or the levels of the nation and the super-nation (an example of the latter would be the European Union); they are designed for each other.

But binding norms seem to become rooted in human beings through Beta, through G. H. Mead's 'Significant Others' (see Mead, 1934), maybe particularly the mother. If Betas now crumble all over the place, down to the nuclear family, even to the mother-child bond, leaving more and more of the raising of children not even to the school where the single class still may have some Beta character, but to the media, parking the children in front of the TV/video, then it would be a miracle if binding and positive norms became internalized. Add to this the well-known content of the media,[24] and the general picture becomes even worse. At least for the 'advanced countries' (meaning advanced both in terms of modernization, including economic growth, and advanced in atomie/anomie), one reasonable hypothesis would be that there is a certain synchrony between the processes of destructuration and deculturation. For other countries there may be important asynchronies[25] to explore.

We would expect a general sense of pessimism to prevail as Beta crumbles, and that is exactly the general finding that emerged from a major comparative ten-nation study, *Images of the World in the Year 2000* (Ornauer et al. 1976): the more economically advanced the country, the more pessimistic in general terms the inhabitants. With Betas crumbling all around them, people may easily become very lonely: add to this the alienation at all levels of Alpha due to the strong rules of substitutability, the exploitation lower down, and the lack of any other moral guidance than individual cost-benefit analysis — how would we expect people to react?

Basically, in the way suggested by the five social phenomena mentioned in the introduction. But the latter already presuppose a weakening of Alpha, not only the alienation and exploitation/repression of formations II and III. Under the conditions of modern society (not yet post-modern), people might react to Alpha as such: if we assume those on top to be basically content, wanting to hang on, then the reaction will mainly come from people lower down. This gives us two formulas: revolt and apathy, boiling and freezing. Who chooses what, both or neither is an interesting problem of social psychology. From a more sociological point of view, these are mass phenomena and solid indicators of malfunctioning, which is not in any way to say that revolts may not be justified. However, if there is something humankind should have learned during the twentieth century, it would be this: a revolution substituting one Alpha for the other, changing priesthood, may not change much.

---

24. Figures often repeated in the US press compare the 15,000 hours of schooling of an average US 18-year-old with 18,000 hours of television viewing, involving an average of one murder per hour, and 340,000 commercials. The point about commercials is not only the idiotizing level of the message, devoid of any intellectual or moral content, but the training in one-way communication: there is no way of talking back, even of asking questions. The only choices are in terms of buying/not buying and viewing/switching off.
25. One hypothesis might be that deculturation is more advanced due to the cultural penetration from the West. This was prepared through evangelical work during centuries of colonialism, to the extent that the recipients were stranded on a cultural dialectic no longer their own, ready to accept the anti-evangelical content of secularism, partly because it took on the same form as secular evangelism. If they are in a state of anomie, but not (yet) of atomie, then we would expect exactly the brutality with which dowry is exacted in some parts of India today. The opposite syndrome, atomie (lots of marginalized, fragmented youths) without anomie (because there is still strong faith in received religion) might predispose for what in the West is called 'fundamentalism'. (The operational meaning of that term is probably any faith that stands in the way of Western penetration.)

Political violence, referred to as 'terrorism', may be a problem of structures/cultures partly of the past. Today the problem may be that there is no structure/culture at all and that violence, hurting and harming, is erupting all over as a consequence of social disorganization. Here is a typology of eight forms of violence:

1. Violence against Nature (ecological crimes)
2. Violence against Self (alcohol/drugs/tobacco, stress, suicide)
3. Violence against Family (child abuse, physical/verbal violence)
4. Violence against Individuals (robbery, assault, rape, homicide)-
5. Violence against Organizations (corruption)
6. Violence against Groups (inter-class, inter-nation violence)
7. Violence against Societies (inter-state violence)
8. Violence against Other Worlds (inter-planetary violence)

Types 3, 4 and 5 are referred to as crimes,[26] and types 6 and 7 as wars. For a peace researcher they are all violence. The arenas differ from one type to the other: all over Nature (as in the rain forests); at home; on the streets; in offices; within a country (internal wars); within the world (external wars); between worlds (so far only as science fiction). But the net result is the same: life is being caused to suffer, being hurt and harmed and traumatized, even ceasing.

All over the world, people are in shock after reading, listening to, viewing, the media. The world seems to be coming apart: each nation wants its state; weapons of all kinds are available everywhere; big blocs are taking shape at the world level; rich countries are set against poor as much as, or more than, ever; the rich in North America, Western Europe and East Asia are pitted against each other; there are new military alliances; culture, and particularly religion, come up against secularized elites capable only of uttering the standard curse: 'fundamentalism'. *Homo homini lupus, bellum omnium contra omnes*; everybody for himself; apparently out of control, unrestrained; disintegration.

One common reflection today is that violence has become more domestic, less global, worldwide. In terms of the above typology that means more violence of types 1–6 (but type 1 is also global!) and less of type 7. It may be too early to judge; the data indicate constancy rather than decline in the level of inter-country violence. There is a perception of decreased threat of a nuclear East–West holocaust in Europe, possibly due to an over-estimation of that danger during the Cold War and an under-estimation of that danger within the Catholic–Protestant/Latin–Germanic vs. Orthodox/ Slav vs. Muslim/Turkish triangle taking shape in Europe. At any rate, with

---

26. Although type 4 is the classical crime, the crime in public space, in the streets, on the roads. What people did inside their homes (type 3) and in organizations (type 5) was once seen as outside the public realm, even as private, to be dealt with by internal justice (meaning the *pater familias* and the chief executive officer, themselves often the offenders).

that danger removed, the world system (in the Northern part of the world) looks to many rather peaceful. Not so domestic society, with nations pitted against each other all over and violence of types 1–5 apparently on the increase in most societies.

The hope of people working for peace has for a long time been to have the world system catch up with the best social systems in controlling violence, for instance by establishing a binding rule-of-law system. The problem, as usually pointed out, is that such rules are not easily internalized in an anarchic system with everybody (meaning the states) out for themselves and nobody really functioning as a Significant Other, a nursing mother.[27] Nor are they easily institutionalized. There are mutual rights and obligations, but if A's right becomes B's obligation and there is no reciprocity, the mechanisms for handling the conflict (the World Court, the Security Council) are imperfect to say the least. Neither rewards nor punishment (positive and negative sanctions) are impressive. What then happens is often hierarchic intervention by big powers.

People may develop all kinds of Beta, not to mention Alpha, structures across borders, but the inter-state structures are thin (this is where anarchy enters) and vertical (this is where hierarchy enters). Is the structure also big? With 184 members of the United Nations and 184 ambassadors, the structure is no larger than many individuals can handle, fitting nicely into their lists of addresses and telephone numbers. Being thin and vertical it could easily become Alpha by adding more members (such as NGOs, or direct relations to the many nations of the world). But it could not easily become Beta. In that case it would more likely be Gamma, with the permanent Security Council members *in loco parentis* of that extended family. Feudal and paternalistic, in other words, and even so the webs of interaction will have to be spun much more densely.

The basic point here, however, is that far from the world system catching up with the better cases of the social systems, it is the other way round: the social systems are 'catching down' with the world system. Read this way, formation IV, replete with atomie and anomie, is a rather adequate image of world society: vertical, with symbolic, abstract relations rather than direct interaction, short on binding norms and altruistic orientation and long on egoistic cost-benefit orientation. There are some Beta structures, as among the Nordic, the European Union and the ASEAN countries. But the formation IV structure is very evident, and the consequence is obvious: instead of efforts at peaceful conflict solution, violence is used, respecting neither common values nor any inner voice of conscience.

---

27. In the self-image of the colonial countries, often referring to themselves as 'mother countries', that may be what they hoped to do. But mothers with that kind of record would hardly serve as moral models at the social level; and at the world level relations are, in addition, remote.

5. FROM NOMADS TO MONADISM: SOME *FORCES MOTRICES*

What follows is not a theory to account for this rather gigantic change in the human condition. Rather than a macro-history, it is simply a catalogue of twelve factors often mentioned in this connection, an annotated list so to speak, even alphabetized to make its atheoretical character more obvious.

Capitalism: The reason why capitalism tends to become not only Alpha but Super-Alpha in its basic structure, even if much is happening within, is the verticality of power that follows when high-quality production factors (nature, labour, capital, technology and management) are monetized, marketed and mobile. As they have to be put together for production, they tend to flow together — or at least to be controlled together — from a Centre.

The Centre uses high-quality factors for high-quality products, in exchange for lower quality factors and products from the Periphery. Capital is supposed to beget more capital, either directly in the finance economy (speculation) or indirectly when invested in production factors used to produce goods and services in the real economy (production). Much begets more, which does not mean that little begets less; the cake may expand, but then often at the expense of the external proletariat, nature and/or future generations. What was new about capitalism was not that the economy had a peaked structure. What was new was the mobility, not only into the Centre by investing some initial capital, much hard work, saving, greed and inconsideration, but also out of the Centre through bankruptcy or lack of dynamism. The result was an anti-feudal revolution. For the continuation, see Socialism.

Democracy: Of course elections are one way of ensuring not only rule by the consent of the ruled, but also nonviolent transition from one set of rulers to the next, if they respect the secret ballot. The problem is the Alpha nature of that type of democracy; a relation between a Centre of contending Rulers and a Periphery of the Ruled, turning the pyramid upside-down once every four years (or so). This Alpha shape of modern democracy, Democracy II, differs from the Beta shape of a more primordial Democracy I: a group (a small company, a small community or a neighbourhood, a family or a group of friends, the elders in a tribe) engaging in dialogue over issues until consensus is obtained. In the latter case, the relation is horizontal, everybody can address everybody's concern, the outcome is unknown in advance, there is neither winner nor loser — and in good dialogues, only winners.

Differentiation: Another term is 'division of labour', seen in a long social philosophy tradition (Adam Smith, Herbert Spencer, Emile Durkheim, Ferdinand Tönnies, Max Weber) as a basic condition for social progress and economic growth in particular. The total human activity called Work is not only subdivided into tasks and sub-tasks, but new tasks are continuously created. As in production of goods and services so also in the production of knowledge: undifferentiated Philosophy is subdivided into disciplines and sub-disciplines that in addition are hyphenated into cross-disciplines. The structure of the sets of tasks and disciplines is highly complex, but the general idea is differentiation, and with it fragmentation, atomization of the individuals having these tasks and disciplines as their job. A book, *Limits to Differentiation*, is crying to be written.

Economic Growth: The process is almost inconceivable without a culture accommodating not only hard work and saving, but also greed and inconsideration. Systems may differ as to whether the pressure is put on the internal or external proletariats; on nature, self or the future. But something has to be moved, or transformed, or both (see Capitalism); and in the process organic relations of people to others — or of raw materials to surrounding nature — will be cut or at least transformed. The open wounds in quarries and mines have their counterparts in the open wounds in souls detached from each other through excessive mobility and transformation. Inconsideration means insensitivity to wounds in Self, Other and Nature. Beta structures break down; partly dehumanized Alpha structures are poor substitutes.

Economism: The term is interpreted here as a state of mind, not to be confused with the economy (the cycles linking Nature, Production and Consumption) or economics (the science about these cycles, today essentially a description and theory of one particular economic system, capitalism, hence a science that more properly should be called 'capitalistics'). Economism, or the culture of *homo economicus*, can be conceived of as a syndrome:

- a focus on material/somatic satisfaction by goods and services;
- a focus on the human individual as the unit to be satisfied;
- a focus on cost-benefit analysis to guide individual choices.

The syndrome not only detaches individuals from each other by making the single individual the supreme decision-maker (egocentrism), but also detaches satisfiers (goods/services) from each other as objects to be possessed and consumed one by one. Costs and benefits are then used to establish preferences. There are severe problems with this syndrome/mind-set:

- in practice only a limited number of satisfiers can be used, by definition excluding the externalities of economic action;
- absolute values (with infinite positive or negative utilities) will be excluded or relativized since they will overrule others;
- individual preferences are not easily reconciled collectively.

Destructuration and deculturation are the — considerable — costs of the breakdown of the holism of the actors in collectivities, the holism of the object-world, and of absolute values. These costs are also built into the technique used: product-sum maximization, which becomes very unwieldy for collectivities of non-harmonized actors and high numbers of satisfiers, and useless for absolutes. The result is even more atomization, destructuration and the deculturation implicit in rejecting absolute values. Thus economism becomes the ultimate consequence of Roman Law.[28]

Gender: Let us pick up just one factor: how the genders seem to differ in their preference for Alpha (male) and Beta (female) and No (male) structures, assuming that women prefer to relate and network — not to be isolated in loneliness, nor to be isolated at the top of a hierarchy. Thus a major force behind the drive toward Alpha-ization, and then toward monadism, from formation I to II, II to III and on to IV, would be patriarchy: the leading structure is the structure of the leading class. That should also apply to culture: male preference for deductive thinking and submission to first principle is compatible with a transcendent God, less compatible with immanence. But that also opens up the possibility for a major therapy: parity instead of patriarchy — provided women have not become clones of men in the process.

Globalization: If this term stands for global mobility of production factors and products, with more standardization of structures and cultures, then the consequence is to speed up the transition into formations III/IV. Larger domains for structures and cultural meanings imply thinner scopes and more reliance on least common denominators, with structural and cultural specifics receding into the background. Given the variety of idioms around the world, super-super-Alphas with truly global reach will be symbolic — based on mathematics, computer language, body language (sport as a universal idiom) and/or on concrete objects, goods, like people with no common idiom pointing and touching. 'Here are no Greeks, no Jews; no women, no men: we are all one in Coca-Cola' is reality, not a bad joke or blasphemy. The same goes for structures: no cohesive Alpha has so far emerged covering 6 billion human beings except one — global television. There are two layers: one sender, billions of receivers. There is no horizontal interaction; they relate via the apex.

Will this structure endure? Probably not: sooner or later it will go the way of all Alphas. Small Beta groups take shape: like guerrillas, they will relate, unite and revolt. The condition is their ability, underestimated by Marx, to overcome structural, cultural and geographical divides. But the global market prophets may have underestimated the fact that in its wake will follow globalized worker/trade union and consumer movements. 'Proletarians (and consumers) all over the world, unite!' may have a reincarnation. Consumer sovereignty, if exercised on a truly global basis, may become a major force at the same time as nation state democracies crumble under the weight of global forces beyond their control.

---

28. See the excellent work by Susan George on the World Bank as a religion (George and Sabelli, 1994). The combination of male, mainstream economist and (probably) Protestant extraction is not very promising.

Actually, globalization may also run into another problem of an equally or more serious nature. Competition has kept capitalism innovative: not only the micro-competition from other firms in the same branch (BMW versus Mercedes) nor the meso-competition from another country (Germany versus the United Kingdom), but the macro-competition from other civilizations with other capitalisms (Buddhist-Confucian versus Judeo-Christian). Globalization will keep the micro and meso challenges but may strive to iron out the macro differences through homogenization into a global business culture. This means a severe reduction of the Toynbee factor of challenge followed by the creative response that presumably keeps minorities in power. And Alpha is, by the very definition built into its pyramidal shape, run by a small minority (relative to the other layers) in need of constant renewal of personnel and ideas. Globalization means mono-culture, less diversity, less symbiosis, less resilience.

Health: The concern for health fits into the general picture of secularization in two important ways: as focus on the body rather than on the state of the mind and the spirit; and the translation of eternal life/salvation into high life expectancy. Of course modern man enjoys lower morbidity and mortality. But there are no gains without some price to be paid, and the price is in the cultural rather than structural sector. Could it be that the healthy body is less able to share the suffering of others at the same time as health — one's own and that of others — is taken for granted, being no source of shared joy either? Could it be that health leads neither to a culture of compassion nor to a culture of submission (following in its wake), but to a culture of egocentrism?

Human Rights: In principle human rights protect exposed individuals, emphasizing the privacy of the individual human body, of the individual human soul/mind/spirit and the equality of all categories of humans relative to the law. Human rights soften relations between the Centre (the state) and a Periphery of individualized citizens, which is good. But the doctrine can also emphasize reliance on a protective, soft Centre rather than human reliance on each other — an ethics of Alpha submission rather than Beta compassion, designed to soften (not weaken) the strong Alphas of formations II and III, in ways leading to formation IV.

Industrialization: There is no doubt this was a major factor in the transition from formation II to III and led to well-known problems of vertical division of labour (exploitation) within the company, between employers and employees; within the country, between raw materials and industrial goods producing districts; and also between countries according to degree of processing. The organization at all three levels was Alpha, with a plethora of Beta groups flourishing at all points, from boys' clubs of employers to workers' collectivities (not the same as Alpha-type trade unions) controlling the level of commitment to the firm. So industrialization has been accompanied by anti-Alpha revolts of all kinds — from sabotage, work slowdowns and company strikes to general strikes, and anti-colonial and anti-neo-colonial movements. The struggle is still on. But the focus here is more on a robotized, automated, symbolic interaction pattern than on industrialization, if that is still the word. From the perspective of destructured and decultured post-modernism, symbolized by robotization, labour and Third World struggles look almost utopian: people still relate to each other.

Literacy: Literacy can only be understood in terms of its alternatives: oralcy on one hand and picturacy on the other. Oralcy has as a necessary condition memory, stored in the brain. Does it not stand to reason that what has to be memorized often is more easily remembered, recalled, related to others in Beta-type relations (rather than the Alpha-type relation of readers to authors) and for that reason may be more compelling? The decalogue can be retrieved from books and computers. But does that have the same binding quality as moral commands committed to the individual memory? If not, is literacy, however precious, not also paving the way from formation II into numbers III and IV?

Picturacy (TV, video) in principle mirrors reality and in practice constitutes a virtual reality, an 'as if' (*als ob, comme si*) reality. The choice has been made for the viewer, as subjectively as any choice. Synchronic perception complements the diachrony of oralcy and literacy, but is also more easily confused with reality 'out there'. This, then, adds to detachment in dehumanized structures and relativized cultures.

Migration: Whatever the reason, massive migration across borders, which often also means across cultures, will considerably speed up transitions to formation IV, even to the new formation I of the Los Angeles metaphor. Thus a person, with or without friends and family, detached from the structures of the country of origin, arrives in the host country, presumably with his/her culture more or less intact. There has to be some attachment to a new Alpha structure, relating to the new State (permits, etc.), the new Capital (job, etc.) and maybe also some footholds in the new Civil Society. However, the cultural idiom will be thin indeed. The host culture is not easily internalized. A likely result is a tightly spun Beta group of immigrants suspected of being predators rather than prey (or, often, both): Los Angeles. This should not be confused with colonialist transfers of total societies, with the host population marginalized or exterminated.

Poverty: Of course, poverty is important as a problem of all formations, when instead of looking at structures and cultures we focus on basic human needs and their satisfaction. But from a structural point of view, poverty does not necessarily lead to atomie. It can also lead to tightly woven Beta groups fighting poverty together.[29] Nor does it have to lead to anomie. It can

---

29. This, of course, is the major theme of the Oscar Lewis tradition in this field, *la cultura de la pobreza*, today often seen as an obstacle to growth because potential entrepreneurs are locked into Beta-type solidarity. The free float upwards by their own buoyancy.

also lead to the famous culture of poverty of the *favela* which may sustain rather than negate poverty, but also make it more bearable. The worst poverty would be needs-deprivation combined with atomie and anomie, in other words the poverty of formation IV. This may very well be the condition under which Los Angeles is no longer a metaphor, but a world reality.

Another, updated perspective on poverty might bring in the jobless growth characteristic of the present world economy. Nevertheless, the distinction between employed and unemployed is too sharp. More typical are under-employment and underpayment, in the sense that the concept of the breadwinner able to feed a whole family ('one job-one family') is disappearing. In principle this should force a number of people, in a family or another kinship unit, or in a neighbourhood or a commune, to join their incomes so that all can live from it, thereby fostering Beta restructuration and solidarity.

Roman Law: It was all pre-programmed in Roman Law, if the following reading of that law or basic philosophy is accepted (see, e.g., Santos Justo, 1994). In what we generally assume to be true about primitive societies, holism is a basic figure of thought, both for Humans and for Nature, although in practice this applies mainly to the near-Humans, meaning the in-group, and the near-Nature, meaning this side of the horizon, which for a nomadic people is considerable. This is very far from a basic figure of thought in Roman Law, which is *dominio* or individual ownership. The ownership or use concept of primitive peoples is a coupling of two holisms: we as a group use, with care, what we find in Nature. To go from that figure of thought to the Roman holisms must yield to atomism. Humans must be subdivided into persons capable of ownership, an example being the *pater familias*, another the emperor; and Nature subdivided into entities capable of being owned, as land, plots and minerals, plants, animals, slaves/women/ children. For this a census gradually had to take shape, and the sciences of geometry, geology, botany and zoology. But once the subdivision was done on both sides of the Humans–Nature divide, with the holisms broken, the totality could be sewn together again, the Roman (to become the Western) way: through a one–one mapping of juridical persons on objects, the *dominio*. What belonged to everybody would then belong to nobody; *res communis, res nullius*. For the non-Western world, ownership was acquired by applying the 'first come, first see, first own' principle, through 'discoveries'. There were transitional formulas: the Emperor is the only juridical person, like the *pater familias* for the family, assuming *dominio* of everything — in the West scorned as 'Oriental despotism'.

Socialism: We know it in its Stalinist and post-Stalinist configuration as super-Alpha with, say, 400 people planning for 400 million in Eastern Europe and the Soviet Union (about the same structure as regional tele-vision). Means of production were collectivized — but not at the level of communes as commune-ism but at the level of the state, as state-ism (*étatisme*). Revolts were inevitable, not only because of the brutality and repression of (post-)Stalinist countries. Planning made people passive, expropriating from them not only the right to plan their own production, but even to plan their own consumption and the economy of their own household, restricting the range of what was available. Then people demanded their right to be subjects of their own economic fate. For the continuation, see Capitalism.

Urbanization: The city is a giant Alpha in administrative and often economic terms. But it is also better suited to host countless rich, diverse, shifting and symbiotic Betas than any other human habitat, if for no other reason than simply because it combines size and proximity into propinquity. It has other problems, such as slum formation and the alienation of those who are marginalized. The young, the old and women are often excluded from the rich Beta variety of bars and clubs. Moreover, modern cities are better designed for cars than for people, eliminating many good meeting places such as parks, open land, old buildings. Like industrialization, urbanization played a key role on the way from formation II to formation III. But even if very dramatic in many places, these are the problems of yesteryear. The problem now and in the future is to ensure that the solutions to these problems do not carry the stamp of formation IV.

6. TWO THESES ON SOCIAL DEVELOPMENT

There will be no surprises for the reader in these two theses on social development as the antidote to social disintegration. The first thesis reads: Create strong Alpha and strong Beta structures, to promote structuration and reverse destructuration. The second thesis reads: Promote immanent and transcendent religion, to promote culturation and to reverse deculturation.

If enacted this would place us in formation II, which has been called 'traditional society', referring to that cycle of human history. But the definitions of these formations transcend the concreteness of the travel from nomads to monadism. We also have a future, and the hunch derived from these deliberations is that we need both Alpha (because 'some big is necessary') and Beta (because 'small is beautiful'). This in no way means moving backward in history (which would be impossible anyhow), but trying to create a new cycle. A not-very-promising beginning has already been indicated: not only the tribal warfare in the wastelands of Los Angeles, but also the warfare in Ulster, ex-Yugoslavia, ex-Soviet Union and Turkey in Europe; Rwanda, Somalia, Liberia and Sierra Leone in Africa; Guatemala and Mexico in Latin America; Myanmar, Indonesia and Cambodia in Asia, to give some examples. Strong on Beta, weak on Alpha — and very violent.

One still-positive example of formation II comes to mind: Japan. Betas in the form of cohorts are incorporated into the Alphas of bureaucracies and corporations by way of lifelong employment (so that people stay together inside the organization) and seniority promotion (so that people stay at the same level for some time, being promoted together at least to start with).[30]

At the same time, Japan also benefits from the co-existence, in one society, of transcendent religion (State Shinto), immanent religion (Folk Shinto, Buddhism) and secularism (Confucianism). In principle, a Japanese person not only lives both in Alpha and in Beta, but may also pay allegiance to all three systems of faith at the same time (and, in addition, to Christianity and Rationalism). Thus we would expect a certain resilience in Japan, being both structurally and culturally intact, playing on both structures and both cultures. This might look like redundancy, but the key to resilience is exactly that, redundancy to be on the safe side. Hence we would expect relatively low disintegration rates of the usual kind, adding divorce to the typology of violence.

Japan is exposed today to tremendous pressure both from the outside, particularly from the United States, and from the inside, maybe particularly from bureaucrats, businessmen and scholars who have been to the United States and found the society liberating. As mentioned above, Beta and immanent religion can be confining; Alpha and transcendent religion both open up grand vistas. But the conclusion from these deliberations is to be very careful: the costs of that type of modernization are enormous and the remedies not very clear, as moving backward, recreating past structures and cultures, may be impossible. To pressure Japan into policies that will have moves toward formation IV as a likely consequence should be classified as some kind of social crime, 'structurocide' *cum* 'culturocide'. This does not mean that Japan and similar countries are perfect. With more emphasis on social growth and costs, and less on economic growth and costs, however, good policies should emerge.

In general, the first thesis would have two sub-theses: to recreate Beta and to rehumanize Alpha. One way of doing this is found throughout Western countries: create Beta inside Alpha of any kind — bureaucratic, corporate, academic. Individualism being so basic to Western cosmology, the Japanese way of tying people to the same organization for life will almost have a taste of imprisonment, and parallel promotion would disregard differences in individual potential and merit. Beta integration does not have to be based on cohorts (i.e., generations); it can also be work-related. The problem with experiments in team work and team teaching would be the scarcity of compelling indicators of the value of social integration when there are few, no or even negative economic gains. At present the significance of social integration must come as a credo.

Thus in any trend to abolish assembly lines in favour of teams assembling a product together, there is a clear potential for some Beta growth and some

---

30. However, the flip side of this 'workship' Beta structure may be an absence of Beta in other contexts, such as neighbourhoods.

Alpha decline. The same applies to modern office landscapes with a high level of mutual visibility, easily organized tasks, grouping together those who should work together. The contrast would be the one person/one office structure, an architectural recipe for fragmentation, with the lunch, the coffee-break and the water-cooler as the only alternatives. These are not so likely to be well suited for production-oriented Betas, with the exception of the 'business lunch'. But what, then, happens to reproduction-oriented Betas?

At the universities this would point to the colloquium as a fine Beta structure, for professors and for students. In the United States these structures are remarkably infrequent. In banking this might point to the interesting lead by the Grameen Bank introduced in Bangladesh. Really poor people do not have equity for bank loans; if they did, the loan might not have been needed. Instead ten persons guarantee one tenth each and together they constitute a Beta group around the debtor.

This reminds us of the famous *Zehnergruppen*, groups of ten people working together, introduced in economic organizations in the former East Germany to increase production and productivity. As such, they may have failed; but as Beta groups they seem to be much missed. Of course, Western capitalist society has much to offer in terms of voluntary organizations (although they often acquire Alpha character, becoming big and formalized), but they are usually not directly work-related. Another interesting Beta innovation is, of course, what in German is called the WG — the *Wohn-gemeinschaft*, the 'commune' of like-minded people living and to some extent consuming together, sharing all the work of the household; an extended family except for the kinship factor. Of course this illustrates a longing for Beta in a society where even nuclear families collapse. It should not be judged by the ability or wish of the members to stay together as 'real' family members; the socio-logic is different. It may also be a major way of internalizing conviviality norms.

In order to rehumanize Alpha, simply ban all automated responses, let people have a chance to put their questions to a human being and get human answers — however fallible — back. The social costs of not doing so will by far outweigh the economic costs of employing more people in the services. Moreover, such positions do not have to be full-time jobs. What is needed is humanity.

Turning to culture, in the narrow sense used here of binding ethical rules: in the choice between an ethics of compassion and an ethics of submission, between a religion of meditation and one of prayer, the answer might be to choose both, with an important proviso to be spelled out shortly. There is much to build on; rich religious experience to draw upon. There is also room for secular approaches, perhaps not the Enlightenment cult of Ratio so much as the general wisdom of 'reciprocal rights and obligations', found all over the world, with at least some ethical inspiration to be derived from its moral basis, the *lex talionis*, both in its negative and positive formulations.

But there is another distinction that may be more important than the sacred-secular and immanent-transcendent: hard vs. soft. The word religion comes from *religare*, to relink, reconnect with that out there, the holy, the sacred. Union of some kind is the goal of all religions, union with Others (past-present-future) through immersing oneself fully in the net of compassion with all life, with God and others in the afterlife, and submitting to His commands. In mysticism this experience probably becomes like a light so strong that everything else loses its contours.

Imagine now a circle around this epicentre of religious experience, divided into sectors for each religion. The notion of religion as linking, connecting, unifying is still there. Religion is not used to draw lines between the adherents of this or that religion, nor between the righteous and the sinners. A religion is seen more like a language, an idiom in which religious experience is expressed. This is the soft circle, perhaps found more in religions of compassion than of submission.

Outside this circle comes the circle for hard religion. The names of the sectors for the religions are the same, but the message changes character. The focus is on what divides rather than on what unifies. Other religions are denounced as pagan, or even worse, as heresy. The sinners are in for very harsh treatment; even hell, the torture chamber of hard religion, is invoked for their afterlife. The righteous (from the right religion) are seen as Chosen Persons in the eyes of God and some nations are often seen as closer to God than others, including the sinners and the non-believers. Naturally, hard religions of that type can be well suited as state religions, mirroring in religious terms the struggle among states in world politics.

So an elaboration of the second thesis for social development would be to promote the softer aspects of the religions and try to demote the harder (harsher) aspects. Thus the most important struggle in the religio-scape — the religious landscape of the world — is not the traditional struggle among religions as to which one is most suited to carry humanity forward, but the inner struggle between the unifying and the divisive forces. 'Soft religionists of the world unite, you have only your harder brothers and sisters to lose'? Not quite, because that would draw a too-hard line between soft and hard. The important point is that the struggle is within rather than between and that each religion has this struggle on its agenda. Moreover, the harder aspects (Inquisition, witch-burning) have no doubt contributed to giving religion a bad name. Quakers and Sufis, Buddhists and Baha'is offer much softer approaches,[31] but none of them would be entirely free from the harder aspects. For humanists this would imply a softening of the line they sometimes draw between themselves and the religionists, following the

---

31. By and large this has not only theological implications, like the Quaker saying, 'there is that of God in everybody', but also sociological implications: a flatter, more horizontal structure. Thus the usual relation between steep hierarchy and violence ordered from the top (who may bless rather than participate in the action) does not obtain.

tradition of eighteenth century Europe. In short, there is a message to everybody in the word that is No. 1 in the vocabulary of the present Dalai Lama: compassion.

Do these two theses add up to the standard conservative message of family and Christianity? No, but that message is not rejected either. 'Family' is taken in a much broader sense, Beta. Moreover, attention is paid to how to soften, humanize, the other major structural type, Alpha. Instead of Christianity we are of course speaking about all religions, sacred and secular (civil), but limited again to the softer aspects. Nevertheless, conservatives have probably diagnosed the present situation better than many liberals/Marxists/greens, by focusing on one structural and one cultural component. People on the left tend to be almost obsessively focused on some kind of Alpha, its proper design, function and structure; its distribution of rights and duties, power and privilege, at the expense of Beta (except for the greens) and culture, ethos. However that may be, the present paper tries to give something to both, perhaps with the strong admonition to the left of taking culture, ethos, religion more seriously, getting out of the habit of seeing them as 'superstructure' or 'opium'. None of this will emerge automatically and in crises people may also turn to the harder aspects of the religions with divisive messages and Alpha organization. But just as we postulate a normal human Beta drive for the small and tight, why not also dare postulate a corresponding religious inclination?

This short excursion into a very uncertain future brings us to the end of this essay. With structural and cultural ties being dissolved, we are in the — some would say absurd — situation that the most modern and economically/technically developed have become, socially-speaking, perhaps the least developed, or 'de-developed'. Obviously, we are not talking about the relative presence of social services (per 1000 inhabitants, etc.) but of something held to be much more basic: structuration and culturation. Social services may be a part of the problem rather than the solution to the extent that they are operated through increasingly dehumanized Alpha structures. With atomie/anomie being the basic social price paid for modernization because people have taken for granted that society is solid and can be drawn upon for any purpose, the more and most developed have suddenly become the less and least developed.

Does that mean that the economically/technically least developed are the socially most developed? Not necessarily. Some very poor Third World countries have been ravaged by unspeakable violence between classes, nations and clans, with the rest of the world often siding with one against the other(s). Some of this violence may be attributable to atomie/anomie and there are signs that it has reached the micro level of social organization, with family members butchering each other — in other words, total violence.

However, much is intact — in Southern and Eastern Europe more than in North-Western Europe, in Central and South America more than in North America. One day the present First World may ask the present Third World for advice about social development. If that happens, the world will have taken a major step forward.

REFERENCES

Childe, V. Gordon (1946) *What Happened in History?* New York: Penguin.

Galtung, Fredrik (ed.) (1994) *Korruption*. Berlin: Lamuv.

Galtung, Johan (1978) *Methodology and Ideology*, Copenhagen: Ejlers.

Galtung, Johan (1979) 'Alfa y Beta y sus muchas combinaciones', in Johan Galtung and Eleonora Masini (eds) *Visiones de sociedades deseables*, pp. 19–95. Mexico: CEESTEM.

Galtung, Johan (1980) 'On the Future of the International System', in Johan Galtung *Essays in Peace Research*, Vol. IV, pp. 615–44. Copenhagen: Ejlers.

George, Susan and Fabrizio Sabelli (1994) *Faith and Credit: The World Bank's Secular Empire*. London: Penguin.

Haase, Dieter (1993) *Mein Name ist Haase*. Celle: VDS Verlag.

Mead, G. H. (1934) *Mind, Self and Society*. Chicago, IL: Chicago University Press.

Ornauer, Helmut, Håkan Wiberg, Andrej Sicinski and Johan Galtung (eds) (1976) *Images of the World in the Year 2000*. The Hague: Mouton.

Redfield, Robert (1953) *The Primitive World and Its Transformations*. Ithaca: Cornell University Press.

Santos Justo, A. (1994) *Fases do desenvolvimento do direto romano*. Coimbra.

Swiss Academy for Development (1995) 'Needed: A New Anomie Concept for Development to Reduce Global Destabilization', Chairman's Report to the World Summit for Social Development (Copenhagen, 6–12 March 1995).

Notes on Contributors

Lord Dahrendorf is Warden of St Antony's College, Oxford, and a Pro-Vice-Chancellor of the University of Oxford. Born in Hamburg, Germany, he has been Professor of Sociology in Hamburg, Tubingen and Konstanz, as well as serving as a Minister of State in the Foreign Office in the German Government. After a period with the Commission of the European Communities, where he was responsible for foreign trade and external relations, and later for research, Lord Dahrendorf went to England; he was Director of the London School of Economics for ten years. His books on social theory and political philosophy — including *Class and Class Conflict*, *The New Liberty*, *Law and Order* and *Reflections on the Revolution in Europe* — have been extremely influential in shaping the debate on citizenship and democracy in the Western world.

Amitai Etzioni is currently president of the American Sociological Association and University Professor at George Washington University, Washington, DC. During the late 1980s, Etzioni became especially well known for his challenge to neoliberal economics, presented in his book *The Moral Dimension: Toward a New Economics*. He has questioned the view that society and politics can be understood as exercises in the individual calcula-tion of cost and benefit. His *Spirit of Community* laid out an alternative agenda for 'communitarianism', a movement which has gained considerable force in the USA in the first half of the 1990s. Amitai Etzioni is the founder and first president of the International Society for the Advancement of Socio-Economics, and editor of the communitarian quarterly *The Responsive Community*. His articles appear in *The New York Times*, *The Washington Post* and *The Wall Street Journal*, and he is a frequent guest on network television programmes in the USA.

Johan Galtung is one of the founders of modern peace research. Turning to the social sciences after initial training as a mathematician, he established the International Peace Research Institute in Oslo in 1959, and was its director for ten years. He was also the founder-director of the *Journal of Peace Research* and of the Inter-University Centre in Dubrovnik, which became a meeting place between East and West. Galtung has served as consultant for a score of United Nations agencies and has held visiting professorships in China, Japan and the USA. He is currently Professor of Peace Studies at the University of Hawaii and at Witten/Herdecke University in Germany. In 1987, he was granted the Right Livelihood Award — often called the 'alternative Nobel Prize' — in the field of development.

Anthony Giddens, Professor of Sociology and Fellow of King's College, Cambridge, is one of the most influential social theorists of the late twentieth century. In his own books and lectures, Giddens has challenged and renewed the sociological tradition; and through his work as editor of journals and monograph series, he has had an even broader impact on the way social scientists think about the contemporary world. At present, he is (among other things) Senior Editor of the journal *Theory and Society*, and Director of Polity Press. Professor Giddens' recent work has focused on the meaning of modernity, explored in *The Consequences of Modernity* (1990), *Modernity and Self-Identity* (1991) and *The Transformation of Intimacy* (1992).

Eric Hobsbawm, one of the century's most distinguished social historians, has helped us to understand the forces which created the modern industrial world. His books on peasant society, the working class, bandits and rebels, revolutionaries and capitalists, industry and empire are classics in the field of Western European history, and have been translated into many languages. Three of his latest books are *Politics for a Rational Left* (1989), *Nations and Nationalism since 1780* (1990) and *Age of Extremes: The Short Twentieth Century* (1994). Eric Hobsbawm is Emeritus Professor at Birkbeck College, University of London, and Emeritus Professor at the New School for Social Research in New York.

Fatema Mernissi, one of the most eloquent voices of feminism in the Moslem world, is Professor of Sociology at the Institut Universitaire de la Recherche Scientifique of the Université Mohammed V, in Rabat, Morocco. For the past decade, she has analysed Islam as a religious text which structures the management of politics and the economy — and defines the roles of men and women — in Moslem societies. Her work has taken the form not only of published books and articles, but also of promoting dialogue: bringing men and women together in public discussions and collective projects to confront some of the most serious issues of livelihood in their own communities. Professor Mernissi has served as a consultant for a number of United Nations agencies. Two of her best-known books are *Beyond the Veil:*

Male-Female Dynamics in a Moslem Society and *Dreams of Trespass*, a beautiful and troubling evocation of her childhood in a harem in Fez.

Tetsuo Najita served as Director of the Center for East Asian Studies of the University of Chicago from 1974 to 1980, and is currently Chair of the Department of History at the same university. From 1991 through 1993, he was Vice President and then President of the Association of Asian Studies, which brings together and represents academics, journalists and other specialists concerned with Asian affairs both in the USA and abroad. Professor Najita has written numerous books and articles on Japanese political and intellectual history, published in Japanese and English. His book on *Hara Kei in the Politics of Compromise* received the prestigious John King Fairbank Prize in East Asian History from the American Historical Association, and another book, *Visions of Virtue in Tokugawa Japan*, was awarded the Yamagata Banto Prize from the Prefecture of Osaka in 1989.

Emma Rothschild is Director of the Centre for History and Economics, and a Fellow of King's College, Cambridge. She is also a Distinguished Fellow of the Center for Population and Development Studies at Harvard University, and Fellow of the Sloan School of Management at MIT. Professor Rothschild has written extensively on economic history and the history of economic thought. In addition, she is deeply involved in probing such contemporary problems as military spending and the impact of technology on the environment. Several organizations draw upon her expertise in these areas, including the Royal Commission on Environmental Pollution (of which she is a Member) and the Governing Board of the Swedish International Institute for Environmental Technology.

Wole Soyinka is a Nigerian writer and lecturer whose plays, novels and poems have been read by millions of people around the world. In trying to select his finest writing, many would choose his autobiographical novel, *Ake: The Years of Childhood*; others would refer to plays like *Madmen and Specialists*, or poetry such as that collected in *Mandela's Earth and Other Poems*. In 1986 Soyinka was awarded the Nobel Prize in Literature; but he has also been recognized for his commitment to democracy and his defence of human rights in Africa. Between 1967 and 1969, he was a political prisoner in Nigeria; and in late 1994 he was forced to flee the country to escape confinement once more.

Tatyana Tolstaya is Associate Professor of Russian Literature at Skidmore College and Visiting Senior Fellow in Slavic Languages and Literatures at Princeton. Her essays and articles on contemporary Russian society and politics have appeared in *The New York Review of Books*, *The Guardian* and *The Times Literary Supplement*, among others. Professor Tolstaya's two books of collected short stories (*On the Golden Porch* and *Sleepwalker in a*

Fog), both published by Alfred Knopf in the USA, have been translated into a dozen languages. A prolific lecturer, she is also on the editorial board of a number of journals, including *Syntaxis* (Paris), *Common Knowledge* (Oxford) and *The New Republic*. She received the Pushkin Prize in 1989 and the Premio Grinzane Cavour in 1990.

INDEX

Abdelwhahab, Mohammed Ben, 39
Abiola, B.M.K., 81
Africa, 22, 85; nations in, 8–9, 74, 78
Al Azmeh, Aziz, 46
Albania, 92
Albert, Michel, 24
Algeria, 68
Ali Abderazik, 46, 47n
Alpha/Beta structures, 173–5, 176–80, 184–7
Anderson, Benedict, 63
Anderson, K., 95
anomie, 14–15, 31, 168, 169, 184–7
Anyaoku, Emeka, 84
Arab world, 6–7, 43; the West and, 44–9, 50, 52
arms industry, 43, 44–5
Asian values, 31–2
atomie, 184–7
authoritarianism, 6, 31–4

Baltic states, 115
Beck, U., 154
Bendix, R., 40n
Bloch, Camille, 127, 128
Brahim, Moulay, 46n
Burke, Edmund, 132
Bush, George, 48

Canada, 87, 98, 100
capitalism, 33, 188
caste systems, 175–6
centralization/decentralization, 36–7, 61, 62, 63
Certeau, M. de, 142
Charity Offices and Workshops, 127
Chechnya, 87, 116
China, 3, 32, 92
Christopher, Warren, 44
citizens, state and, 56, 57, 58, 61; see also civil society
civil society, 20, 21–2, 25; globalization and, 27–31; state and, 10
civil war, 5
communitarianism, 10, 30, 90–1, 91–8; cultural oppression in, 96–8; majoritarian, 95–6; values of, 97–8; competition, 6, 7
community, 58, 63, 93–4; cultural oppression
competitiveness, 34, 37
Condorcet, M.J., 12, 120, 128–31, 135, 136, 137
consumption, by the poor, 123
contract co-operatives, 143–9
co-operatives, in Japan, 143–9

Copenhagen Conference 1995, African ministers' presentation to, 69–70, 82–3
corruption, 156, 170
cosmopolitanism, 142
credit associations, 147, 148
Crittenden, Jack, 93
cultural oppression, 96–8
cultural transformations, 180–3
culture, 166n, 168, 196

Dahrendorf, R., 5–6, 11, 42
D'Antonio, Michael, 96
Dasgupta, Partha, 35
deculturation, 185, 189–90
Dekmegian, Richard, 48
democracy, 10, 15, 23, 188; in Arab world, 49; co-operatives and, 149; threat to, 98–100
Denmark, 148
Depots of Mendicity, 127
Derber, Charles, 95
destructuration, 189–90
detraditionalization, 154
development, 21, 22, 68–9, 70; and nationhood, 71, 72; see also social development
dictatorship, 85–6
differentiation, 188–9
Dostoevsky, F., 117
Douglass, R.B., 94

Eastern Europe, 3, 23
ecology, 161
economic growth, 189
economic opportunity, 20, 23
economism, 11, 23, 35, 189
ecosystem, 165
education 36, 124–5, 129–30, 132, 136
Egypt, feminist movement in, 47
emancipatory politics, 157–8
employment, 2–3, 29, 36, 160–1
Enlightenment, 125, 151, 154, 182
Ensor, George, 134
Estonia, 115
ethics, 13
Ethiopia, 73
ethnic groups, 98–100
ethnic violence, 5, 9
Etzioni, Amitai, 10
European Union, 65, 74, 174

Field, Frank, 36
First World, 19–23
flexibility, 25, 29–30
Fowler, R.B., 93, 94